Weather: Spaces, Mobilities and Affects

This book delves into the everyday spaces, diverse mobilities and affective potency of weather. It presents cutting-edge research into the multiplicity of weather phenomena and analyses the lived experiences of humans in conjunction with contemporary issues, notably climate change.

The book considers how everyday experiences of weather in the mundane lives of people are linked to broader changes in weather patterns and climate change. Heat, dust, ice, snow, precipitation, sunlight, clouds, tides and fog are states of weather that impact on the ways in which humans become intertwined with landscapes. Our experiences with weather are diverse and ever-changing, and engaging with weather entangles humans with mobilities, materials and landscapes. This book thus explores affective and sensory resonances, drawing upon a variety of theoretical, empirical and creative material to investigate how weather is perceived in different social and cultural contexts. Key themes focus on the mobilities generated by weather, the affective and sensual potency of weather, and the diverse cultural forms and practices that exemplify how weather is historically, geographically and artistically represented.

Offering a social and cultural understanding of weather events, this book contributes to a growing literature on weather across various disciplines, including human geography and cultural geography, and will thus appeal to students and scholars of geography, sociology, humanities, cultural studies and the arts.

Kaya Barry is an artist and cultural geographer working in the areas of mobilities, migration, tourism, material cultures and arts research. She is a Postdoctoral Research Fellow at Griffith University, Australia, exploring how migration experiences are conditioned through materiality, everyday routines and visual aesthetics.

Maria Borovnik is a Senior Lecturer in Development Studies at Massey University, New Zealand, co-coordinates the Mobilities Network for Aotearoa New Zealand, is on the Editorial Board of *Transfers: Interdisciplinary Journal of Mobility Studies* and is Book Review Editor of the *New Zealand Geographer*.

Tim Edensor is Professor of Human Geography at Manchester Metropolitan University, UK. He has written books on tourism at the Taj Mahal (1998), national identity and everyday life (2002), industrial ruins (2005), light and dark (2017) and urban materiality (2020). He is co-editor of the *Routledge Handbook of Place*.

Routledge Planetary Spaces Series

Series Editor: *Professor Kimberley Peters, Helmholtz Institute for Functional Marine Biodiversity, University of Oldenburg, Germany*

This book series captures the emergent, expanding and important set of discussions on planetary spaces. It provides a home for cross-disciplinary and cutting-edge work that is attentive to the relations between planetary and extra planetary processes and human and more-than-human life.

In a world of uncertainty, flux and transformation, the 'planetary' is becoming an ever-important framework for making sense of socio-cultural, economic, political and environmental change. As it becomes increasingly challenging to draw neat lines around economic crisis, environmental degradation, the reach of the urban and geopolitical fall-out, the planetary provides a means of understanding the intersecting role of 'worldly', 'earthly' and 'more-than-earthly' matter and processes in relation to human and more-than-human life.

We invite scholars from across the social sciences and humanities to publish their original and innovative work in a series that intends to be the 'go to' place for insights concerning planetary and extra-planetary spaces.

Weather: Spaces, Mobilities and Affects
Edited by Kaya Barry, Maria Borovnik and Tim Edensor

Weather: Spaces, Mobilities and Affects

Edited by
Kaya Barry,
Maria Borovnik
and Tim Edensor

Routledge
Taylor & Francis Group

LONDON AND NEW YORK

First published 2021
by Routledge
2 Park Square, Milton Park, Abingdon, Oxon OX14 4RN

and by Routledge
52 Vanderbilt Avenue, New York, NY 10017

Routledge is an imprint of the Taylor & Francis Group, an informa business

British Library Cataloguing-in-Publication Data
A catalogue record for this book is available from the British Library

Library of Congress Cataloging-in-Publication Data
A catalog record has been requested for this book

ISBN: 978-0-367-40639-4 (hbk)
ISBN: 978-0-367-80819-8 (ebk)

Typeset in Times New Roman
by codeMantra

Contents

Figures

Contributors

Gail Adams-Hutcheson is a socio-cultural geographer and teaching fellow at the University of Waikato, Aotearoa New Zealand. Her work encompasses social justice aspects of mobilised communities in post-disaster and farming spaces and places. Adams-Hutcheson's work is qualitative and in-depth and draws on feminist-inspired methodologies. Her latest article 'Farming in the troposphere' examines weather, affective atmospheres and empirical research moments within a farming community in the Waikato region of Aotearoa. She is currently a co-convenor of the New Zealand Geographical Society's Mobilities in Geography Research and Study Group.

Becky Alexis-Martin is a lecturer of cultural and computational geography at Manchester Metropolitan University, UK. Her PhD combined spatiotemporal atmospheric and population modelling and is entitled *RAD-POP: A new modelling framework for radiation protection*. Her current research explores the digital, social and cultural dimensions of existential threats, with a focus on nuclear geographies. In addition to publication in journals including *Geography Compass* and *Eurasian Geography and Economics*, her work has been featured by *The Guardian, The Independent* and *The Conversation*. Dr Alexis-Martin is the author of *Disarming Doomsday*, which has been shortlisted for the L.H.M. Ling Outstanding First Book Prize.

Kaya Barry is an artist and cultural geographer working in the areas of mobilities, migration, tourism, material cultures and arts research. She is a postdoctoral research fellow at Griffith University, Australia, exploring how migration experiences are conditioned through materiality, everyday routines and visual aesthetics. Her creative arts practice informs both the methods and focus of research, with a keen interest in how mobilities and geographical research intersect with creative, participatory and community-oriented forms of engagement. Publications include *Creative Measures of the Anthropocene: Art, Mobilities and Participatory Geographies* (Barry & Keane, 2020, Palgrave) and *Everyday Practices of Tourism Mobilities* (2017, Routledge).

Maria Borovnik is a geographer and Senior Lecturer in Development Studies at Massey University, New Zealand, working in the intersection of mobilities, migration, and development, with a special interest in mobile occupations. Her interests have increasingly involved non-representational, everyday perspectives. She co-coordinates the Mobilities Network for Aotearoa New Zealand, and the New Zealand Geographical Society's Mobilities in Geography Research and Study Group, is a member on the Editorial Board of *Transfers: Interdisciplinary Journal of Mobility Studies;* and Book Review Editor of the *New Zealand Geographer.*

Susannah Clement is a human geographer whose research interests are inspired by feminist thought and the gendered politics of everyday family life. She completed her PhD in 2018 with the School of Geography and Sustainable Communities, University of Wollongong, Australia. Her thesis explored the everyday walking experiences and practices of families living in Wollongong. Clement has published in the peer-reviewed journals *Gender, Place and Culture* and *Children's Geographies*. She is an honorary associate fellow with the Australian Centre for Culture, Environment, Society and Space, University of Wollongong.

Beth Cullen is a postdoctoral research fellow in the School of Architecture and Cities at the University of Westminster in London, UK. She is currently working on the interdisciplinary ERC-funded project, Monsoon Assemblages, exploring relations between changing monsoon climates and rapid urbanisation in three South Asian cities. Her research interests include environmental anthropology, more-than-human ethnography and transdisciplinary approaches for understanding and working with complex socionatural systems.

Faith Curtis is a lecturer in the Discipline of Geography and Environmental Studies at the University of Newcastle, Australia. Her PhD research focused on the way care is performed with people from refugee backgrounds in Australia. It reflected on the relational nature of care performances and offered a more hopeful account of the politics of caring with people from refugee backgrounds. Curtis is also a community development practitioner facilitating collaborations between artists and communities based on a community's desire to achieve artistic and social outcomes.

Lara Daley is a Lecturer in the Discipline of Geography and Environmental Studies at the University of Newcastle and a member of Yandaarra, a Gumbaynggirr and non-Gumbaynggirr research collective working to better understand Gumbaynggirr-led caring for County and agreement-making. Her PhD research drew on the 2014 protests against the G20 Leaders' Meeting in Meanjin/Brisbane to rethink urban anti-capitalist politics and the city in relation to Indigenous Struggles.

Tim Edensor is Professor of Human Geography at Manchester Metropolitan University, UK, and a research fellow in Geography at Melbourne

University. He is the author of *Tourists at the Taj* (1998), *National Identity, Popular Culture and Everyday Life* (2002), *Industrial Ruins: Space, Aesthetics and Materiality* (2005), *From Light to Dark: Daylight, Illumination and Gloom* (2017) and *Stone: Stories of Urban Materiality* (2020). He is also the editor of *Geographies of Rhythm* (2010), and co-editor of *The Routledge Handbook of Place* (2020) and *Rethinking Darkness: Cultures, Histories, Practices* (2020).

Szilvia Gyimóthy is an associate professor in Tourism Marketing at Copenhagen Business School, Denmark. Her research is focused on how tourism is shaping places and place-making practices in the wake of global mobility and mediatised travel. She is an interdisciplinary scholar, bridging across the fields of market communication and branding, tourism geography, and consumer culture studies. Her research projects explored the commodification of rural places and regions along popular consumer culture trends, including the Nordic terroir, adventure sports, Bollywood films and contemporary pilgrimage. More recently, Gyimóthy has been working with sharing economy phenomena, and the marketisation of social relationships in communitarian businesses.

Katharine Haynes is a research fellow in the Centre for Environmental Risk Management of Bushfires at the University of Wollongong, Australia, and at the New South Wales Bushfire Risk Management Research Hub, Australia. Her research explores the human dimensions of environmental change with a focus on community and youth-centred disaster risk reduction and climate change adaptation. Haynes has a special interest in participatory processes and action research as a means for understanding and enhancing community-based initiatives and development. She has conducted in-depth interviews with residents and emergency services following a number of Australian bushfires, floods and heatwave events.

Gareth Hoskins is a senior lecturer in Human Geography at Aberystwyth University, UK, where he teaches and researches on broad themes within social and cultural geography. His previous projects include the ESRC-funded doctoral research at Angel Island Immigration Station, San Francisco, AHRC-funded work on the environmental politics of industrial heritage in California and Kimberley South Africa, and AHRC-funded work on cultural values of historic preservation in Washington, DC.

Ole B. Jensen is professor of Urban Theory at the Department of Architecture, Design and Media Technology, Aalborg University, Denmark. He is deputy director and co-founder of the Centre for Mobilities and Urban Studies (C-MUS). He is the author of *Staging Mobilities* (Routledge, 2013) and *Designing Mobilities* (Aalborg University Press, 2014); the editor of the four-volume collection *Mobilities* (Routledge, 2015); co-author (with Ditte Bendix Lanng) of *Mobilities Design: Urban Designs for Mobile*

Situations (Routledge, 2017); and co-editor (with Claus Lassen, Ida S.G. Larsen, Malene Freudendal-Pedersen and Vincent Kaufman) of the *Routledge Handbook of Urban Mobilities* (Routledge, 2020).

Martin Trandberg Jensen is an associate professor in the Department of Culture and Learning at Aalborg University, Denmark. He is an eclectic theorist interested in tourism, mobilities and the development of new qualitative methodologies. His research draws insights from post-structural theories deriving from human geography, sociology, phenomenology and anthropology. His work often relates to questions of space/place, materiality and affect in everyday life. His work has been published in journals such as *Annals of Tourism Research, Mobilities, Tourist Studies* and *Current Issues in Tourism.*

Jonas Larsen is a professor in mobility and urban studies at Roskilde University, Denmark and PhD-programme leader for Society, Space and Technology. He has a long-standing interest in tourist photography, tourism and mobility more broadly. More recently, he has written extensively about urban cycling and running. He is currently writing a book on urban marathons and is the key academic expert in a new huge research project on urban walkability. His work is translated into Chinese (both in China and Taiwan), Japanese, Polish, Portuguese, Czech, Korean (in process) and Turkish (in process), and he is on the editorial board of *Mobilities, Tourist Studies* and *Photographies.*

Kimberley Peters is based at the Helmholtz Institute for Functional Marine Biodiversity (HIFMB) linked with the University of Oldenburg and Alfred Wegener Institute (AWI), Germany. Trained as a human geographer, she is interested in regimes of environmental and planetary governance. Her work largely focuses on the sea but is now also lifting off to space.

Tonya Rooney is a senior lecturer in early childhood and environmental education at the Australian Catholic University. Her research focuses on children's relations with space, time and more-than-human worlds in contemporary society, and more recently looks into the scenario of Anthropogenic climate change and what this means for young children's lives and futures.

Matalena Tofa is a lecturer in the Professional and Community Engagement (PACE) program at Macquarie University, Sydney, Australia. Her research interests include hazards, disaster risk reduction and communities, community development and Indigenous geographies. She is particularly interested in collaborative and participatory research methods. Tofa has conducted post-event research following floods, bushfires and heatwaves in Australia.

April Vannini is a non-disciplinary researcher and artist living on Gabriola Island, BC. She earned her research degrees in anthropology (MA) and media communication and philosophy (PhD). As an artist, Vannini's individual and collaborative creative process uses a variety of media including photography, video, audio field recordings, multimedia platforms, and found materials for eco-diagramming projects. As a researcher, she has written, published and collaborated on projects as diverse as the Canadian Indian Act, BC Ferries and small Island communities, movement and mobility, and genetic testing of the female sporting body. She is the co-author of a Routledge book with Phillip Vannini called *Wilderness*, 2016 that explores wilderness geographies.

Phillip Vannini is an ethnographer, filmmaker and author who has conducted research on BC Ferries, off-grid living, small island cultures and communities, wildness and wilderness, everyday life, the cultural aspects of the human senses, food and culture, and sense of place. Currently, with April Vannini, he is researching the cultural dimensions of UNESCO World Heritage natural sites, across both Canada and the world. His latest ethnography was published as *Off the Grid: Re-Assembling Domestic Life*, the culmination of three years of research into the lives of people across Canada who live off the grid, a project that led to the production of the film *Life off-grid*. His latest book is *Doing Public Ethnography*, a practical, "how-to" reflection on the methodologies behind public ethnography.

Joshua Whittaker is a research fellow with the Centre for Environmental Risk Management of Bushfires at the University of Wollongong, Australia. His research focuses on the vulnerability and resilience of human communities to bushfire. He specialises in post-bushfire research and has undertaken hundreds of interviews with people affected by bushfires. He has led studies into a number of significant fires including the 2009 Black Saturday fires, the 2017 Sir Ivan fire, the 2018 Tathra fire, and the 2019–20 Black Summer fires.

Sarah Wright is a professor of geography and critical development studies and Future Fellow at the University of Newcastle in Australia. Her research focus is on the geographies of food (particularly working with networks of small-scale farmers and Lumads in the Philippines) and in Indigenous geographies. She is part of the Yandaarra Collective, led by Gumbaynggirr Elder Aunty Shaa Smith, which aims to shift camp together towards Gumbaynggirr-led, decolonising ways of caring for and as Country. She is also part of the Bawaka Collective and has had the privilege of working/living/loving with Yolngu grandmothers and other families for over 12 years.

1 Introduction

Placing weather

Tim Edensor, Kaya Barry and Maria Borovnik

Introduction

Heat, dust, ice, snow, precipitation, sunlight, clouds, tides, fire, ash, haze, fog, particles, high or low pressure, cyclones – these and other weather conditions alter the ways in which humans are intertwined with the world. As annual global temperature records continue to climb and 'extreme' weather makes international headline news, public attention increasingly focuses on the potent effects of weather on our everyday lives and livelihoods. Simultaneously, public debates about the impacts of chemical emissions on changing weather remain unresolved. Yet, although these 'weather-worlds' can be disastrous, disrupting lives, they are also productive of everyday experiences in which kinaesthetic, visual and affective resonances continuously emerge. This book foregrounds the everyday spaces, mobilities and affects of weather, while also registering the recent dramatic changes in weather induced by climate change. We have sought to include a range of cultural, historical and postcolonial aspects in exploring weather from a multitude of viewpoints, and by looking at diverse effects on humans, nonhumans, landscapes and materialities across space and time. In this introductory chapter, we explore five key themes that resonate through the chapters in this book: the place-oriented ways in which weather is experienced and conceived; the diverse mobilities of weather and those induced by weather; the relationship between weather and materialities; the affective potency of weather; and the enfolding of weather into cultural representations. We conclude by readdressing the spatial scales at which weather is understood and apprehended.

Knowing weather, knowing place

We live in places that are ceaselessly assailed by weather, whether this is perceived as consistent and predictable or ever-changing. Weather is a condition of our being-in-the-world, an existential accompaniment to living in place. Along with everyday vegetation, the soundscape of birdsong or traffic and music, local ways of moving and talking, the taste and smell of local food, colours and architectural forms and places of gathering, weather constitutes part of a serially encountered realm, saturated with historical resonances, and contributing to a sense of belonging to place. As we inhabit these weather-worlds, we perform everyday habits that respond to regular patterns of rain, cold and heat, often unreflexively managing daily routines that accommodate the conditions through which we move, work and play.

In carrying out everyday tasks and ordinary practices, people habitually sense place and move through it, for the most part, without thinking, while possessing a competence borne of repeated practice. Weather is thus part of a 'lay geographical knowledge' (Crouch, 1999) that informs how and when particular practices are carried out. Such individual habitual competencies and routine engagements with familiar space intersect with those of others. Through sharing inhabitation in weather-worlds, a sense of 'cultural community' is co-produced by 'people together tackling the world around them with familiar manoeuvres' (Frykman and Löfgren, 1996: 10–11), strengthening affective and cognitive links. Accordingly, as Mick Hulme (2017: 27) insists, all such weather knowledge and practice 'cannot exist separately from the cultures in which it is made or through which it is expressed'. Critically, as Tim Ingold stresses, and is echoed by contributors to this book, weather 'is not to make external, tactile contact with our surroundings but to mingle with them' (Ingold, 2007: 19). Weather gets under the skin of our porous bodies: sunlight tans us, rain soaks us and we are enveloped in fog. And the everyday practices that support our bodies shape weather: transport fumes create haze and pollutants, aeroplanes alter wind corridors, and the food we eat involves intensive water and soil use that exacerbates drought and fire conditions. Awareness of weather conditions initiates anticipation, planning and practice.

This ongoing relationship is underpinned by recent research that shows that 94% of Britons admit to having discussed weather in the past six hours (Maloney, 2017: ix), commenting on the qualities that persist at any time and shape their activities. This everyday weather talk is thoroughly embedded in English language phrases: 'I feel a bit under the weather'; 'it was storm in a teacup'; 'she stole my thunder'; 'he was a fair weather friend', 'I put the wind up him' and 'I'll take a rain check'. And besides being grounded in everyday conversations, weather-oriented habits become institutionalized through scheduled weather forecasts – television and radio bulletins organized hourly, daily and weekly that foretell of coming conditions via established visual and narrative media forms. Weather data is accessed over 9.5 billion times each day (AccuWeather, 2015), delivering localized forecast information to people using smart devices around the world. Part of daily routines, these predictive broadcasts help us to decide on imminent and future courses of action: whether to hold the children's birthday party outside, walk to the shop now or later, wear a thick coat or light T-shirt. Weather forecasts further inform routinized practices at different scales in determining when crops are planted, yachters set sail and holiday seasons are demarcated.

Weather forecasting is itself expressive of a large, routinized operation installed to produce daily schedules organized to collect, analyse and disseminate forms of information. The emergence of the science of meteorology in the 19th century was primarily initiated to minimize loss of life at sea and develop enhanced measuring devices such as anemometers, thermometers, rain gauges and barometers to more accurately capture weather patterns over time. Since these times, technological advances have forged a vast, organized infrastructure that has progressively enrolled radar and satellite technologies, a systematic and bureaucratic infrastructure productive of routinized engagement with the weather. Analysis of weather has become a global endeavour, reflected in transnational systems of forecasting. Globally there are over 10,000 weather stations collecting surface-level data, 1,000 in the upper-air, more than 7,000 ships and 1,000 drifting buoys in the oceans and over 3,000 commercial aircraft measuring 'key parameters of the atmosphere, land and ocean surface every day' (WMO, 2020). Importantly, daily measuring is critical in understanding human action and impact as weather data becomes indisputable evidence of anthropogenic climate change. Accordingly, weather is a global phenomenon that is sensed locally. When checking the forecast on our phone, radio or national meteorology website, we momentarily glimpse the international network of weather that connects places near and far. Increasingly, these 'global' aspects of weather have become part of popular cultural consciousness.

Although weather pervades everyday experience, in contemporary times a battery of technologies and managed environments keeps it at bay and moderates its force. As Vannini and Austin (2020: 1) suggest, we undertake practices 'through which we control, modify, endure, adapt to, enjoy, or remove ourselves from the weather-places we inhabit'. Most obviously, the donning of wellington boots, hats, thick coats and many other forms of clothing is part of an everyday adaptation to weather, and these are accompanied by the shielding effects of umbrellas and parasols. These efforts to ameliorate the effects of weather on bodies have been intensified by climatically controlled indoor environments. For those who can afford it, central heating spreads warmth across domestic space in wintry months, with patio heaters extending heat to the outside realms of homes, bars and restaurants. Air conditioning cools air to minimize the sweat and discomfort posed by heat. As a consequence of these highly managed environments, for many in the Global North, the increasingly sequestered lifestyles of interior environments insulate people from the effects of the weather (Oppermann et al., 2018). The recent pandemic lockdowns have drawn stark attention to the inequalities in experiencing weather. We have seen news about many who are unable to seek shelter or are dealing with inadequate lodgings (Ismail, 2020; Levin, 2020). Keeping out of the weather and 'sheltering in place' in comfortable climate-controlled interiors is primarily for those in the middle-class and white-collar professions. Russell Hitchings (2010: 282) explores how legal professionals working in an office in London were often unaware of seasonal change, so untroubled by the outside weather that they felt ill-equipped to negotiate a sudden cold spell in the weather, as the

> smart shoes he habitually wore could not get sufficient grip onto the icy pavement beneath them and a professional persona was suddenly undermined by the necessity of grabbing onto railings and whatever else that was immediately available.

Hitchings' example contrasts with Susannah Clement's exploration in Chapter 4 of the practical, nuanced ways in which mothers with children negotiate the occasionally rainy conditions of Wollongong, Australia. Clement highlights that when families decide whether to face the rain or shine outside, they negotiate external conditions by selecting appropriate clothing materials and articulate discourses of comfort wherein weather conditions are perceived as invigorating or uncomfortable.

Perhaps because of these routines instilled in daily life that inure people to variations and extreme weather events, and thereby disorient our 'local' weather knowledge, we are yet to adapt to climate changes to come. Yet increasingly, occasions are emerging in which everyday apprehensions of weather in place are dramatized through the acceleration of human-induced climate and environmental change. The advent of increasing incidents of strange weather in local contexts is allied to a growing awareness of floods, bushfires, heatwaves, droughts and tornadoes to consolidate a sense that global climate patterns are awry. Thus 'climate processes are imbricated with ecological and human processes' (Clifford and Travis, 2018: 8) that exceed the abstractions of science. As Katharine Haynes, Matalena Tofa and Joshua Whittaker exemplify in Chapter 12, a small Australian town, formerly conceived by its inhabitants as a safe idyll, undergoes a total transformation as a bushfire fuelled by strong winds sweeps through it. The sensory stimulation that they capture in their interviews with local residents points towards the pervasive, emotional and embodied experiences of weather. Descriptions that account for the effects of the powerful wind on bodies and the heat caught in their throats evoke chaotic, immersive scenes and the stark reality of these extreme weather events. The scent of ash and smoke in the air left a community anticipating their fate, foretelling of evacuations and upheaval. These extreme weather events were watched by the world as wildfires raged in the Amazon, Indonesia and Siberia, and an excruciating eight months of bushfires in Australia destroyed over 19 million hectares in 2019–2020, setting an unprecedented backdrop for the 'new normal' of living with unpredictable weather. In fact, the Australian bushfires were so large they created their own micro-weather systems (Badlan, 2019), producing storms, lightning and tornadoes. For over half of the Australian population, a long spell of poor air quality and the constant scent of smoke pervaded daily experience, and everyday routines were transformed through the constant checking of weather forecasts, watching for emergency services advice, the wearing of face masks and, for many communities, planning for imminent disruption.

Exceptional weather is nothing new, however, and frequently offers opportunities for overturning usual everyday routines and schedules, while also providing memories of times when quotidian habits were suspended. Folk memories endure of London's Frost Fairs, staged when the River Thames froze, providing an occasion for trade, drinking and eating, music and dance on the frozen ice. Similarly, uncanny and rare episodes in which frogs and fish descend from the sky having been swept up by strong updraughts during a storm are part of the folk memories of particular places. We are seeing an echo of these folk tales as long-term weather events disrupt ecological balances: droughts produce conditions for mass fish deaths, and heat waves are so powerful that birds drop dead from the trees in which they shelter. While in the past, such extraordinary weather may have surprised, entertained or unsettled communities, it now urges us to forge stronger connections with the environments that we inhabit. This connection is notably and continuingly strong in Indigenous communities, as can be seen in the declarations of the Rights of Nature. Acknowledging the value of Indigenous knowledge, Ecuador included the right of Pachamāma (Mother Earth) in its constitution and in other contexts, rivers such as the Whanganui River in New Zealand, or the Yamuna River in India, have received a legal right for 'personhood' (Youatt, 2017). The Ngāi Tūhoe in New Zealand are the children of mist and *Hine-pukoho-rangi*, the Mist Maiden, is their ancestor from whom their iwi (tribe) has originated. In this book we foreground a dynamic intermingling between bodies and weather, including nonhuman entities: trees, plants, rocks, bricks, birds, water, dolphins, ships and aeroplanes.

In Chapter 14, Sarah Wright, Lara Daley and Faith Curtis highlight how weather has been long understood as agential within many Indigenous cosmologies. They, for example, draw from the work of Warlpiri man Wanta Steve Jampijinpa Patrick (2015), they explain how 'the cloud that's formed after the hot air and cold air meet and interact' leads to new understandings. Wright, Daley and Curtis explain that 'Indigenous peoples [are] continuing to know and live with weather in diverse and placed-based ways'. Since the colonial establishment of Australian governance and capital, they contend, 'place has been weathered in deeply racist, entitled and possessive ways across time and space'. Racist notions of weather throughout Australian colonial history have played a strong role in choosing the capital. Aboriginal resistance against (post)colonial normative perceptions is at the core of Chapter 14, with the authors advocating for a movement beyond dominant Western narratives of weather.

Affecting both Indigenous and non-indigenous peoples, everyday modes of inhabiting place are traumatically upturned by climatic events. Such awareness, compounded by the connections with increasingly volatile weather in other places, chimes with local conditions. Mick Hulme (2017) contends that such knowledge is based on place-specific engagements with weather, acquired through quotidian pursuits including farming, gardening and other outdoor activities, practices through which local weather conditions signify wider global climatic change. As Hulme argues, 'experiences of weather evolve into knowledge about climate' (2017: 32), knowledge that forms a basis for plans and investments, especially for farmers (Geoghegan and Leyson, 2012) and others who must negotiate both weather events and seasonal swings in natural resources, taking account of 'emergent and fluid' dwelling processes that demand constant adjustments. In Chapter 15, Gail Adams-Hutcheson notes that New Zealand farmers are observant of local weather changes that reveal changes in climate that will increase average temperatures by 1.5°C, with profound effects, with one farmer talking of his feeling that there is 'no escape' because of climate change.

Weather mobilities

Although weather is invariably placed, it is continuously mobile, sometimes rapidly changing from hot to cold or dramatically producing strong winds while at other times, slower changes devolve, as with the creeping advent of gradually colder temperatures. In swirling across space and through place, weather may provoke or hinder mobilities amongst humans and nonhumans. Moreover, weather is a medium that moves material forms across space and equally, flows of weather are diverted by material entities in landscapes. Accordingly, weather shapes the ways in which we move. Whether we seek shelter from the sun or rain, walking on the street side that protects us from the wind, hurrying along when it is cold or lingering in balmy weather, whether we cycle, drive, sail or fly, our activities are rarely detached from choreographies shaped by weather. Walking in rain, as Susannah Clement explains in Chapter 4, requires careful planning, a mindset that is willing and good clothing.

At other times, our movement in weather can be more problematic, as when thick snow or ice covers the ground and coerces bodies to move gingerly. The ordinary routine journeys or 'place ballets' (Seamon, 1979) that we compose as we move across familiar spaces are thoroughly disrupted by such weather events. Yet, the familiar knowledge of a place over time is apt to produce an embodied and habitual experience of walking with it; what Marie Madzak (2020) calls 'moving in sync' with weather also means moving around it, away from it or opening up to it as part of a sensuous engagement. The same is the case for mobilities shaped by forms of transport. A strong headwind may deter cyclists from going out for a jaunt, while stormy seas prevent boats from taking to the water; on the other hand, speedy sailing is facilitated as beneficent winds catch sails and propel yachts across water surfaces. Yet different bodies are able to negotiate weather in different ways according to their experience, physical ability and fitness, age, gender, different styles and acquired embodied skills (Vannini and Austin, 2020). In Chapter 5, as Jonas Larsen and Ole Jensen reveal, runners come to be adept at managing training in the rain, just as experienced skiers learn to move through snow, and drivers learn to drive on icy roads. As Larsen and Jensen explain, 'weather is not only sensed but impacts directly on runners' somatic sensations, internal senses, and for how long they can sustain a certain pace'.

Trips to the countryside, beach holidays and forms of adventure tourism seek to satisfy longings to become more immersed in the external world, to recreate the embodied encounters with weather that once suffused daily existence. These leisurely mobilities in the search for 'good weather' have been the selling point for many tourist destinations and the emergence of contemporary travel cultures. Indeed, this division between times exposed to the effects of weather and sequestration from such impacts might be considered, following Jacques Rancière (2009), as constituting a broader tendency through which a temporal and spatial redistribution of the senses has progressively devolved. Diverging from the familiarity of the daily local weather, in Chapter 2, Phillip and April Vannini discuss how weather is with us at all times, even if we are not necessarily assailed by its impacts. In moving away from the cloistered realm of the university, they illustrate how academic fieldwork may become entangled with place-based weather that has to be negotiated in situ. They highlight how, through fieldwork, they have experienced storms in rough Japanese seas, incessant winds in Patagonia and Iceland, and strength-sapping heat in Tasmania, as well as the conversely enjoyable heat in Belize. In the absence of grounded, routine weather knowledge, they decided to adapt to the conditions that pertained and extended their environmental knowledge.

What might impede the mobility of novices can be readily negotiated by those used to moving in inclement conditions. In addition, mobilities through diverse forms of weather are facilitated by a range of technologies. Fast and long-distance 'hyper' mobility has been enabled by aeroplane technology that has developed to accommodate sudden turbulences in the air, as Kaya Barry's study shows in Chapter 6. She explains that 'sudden changes in an aircraft's pitch or altitude that cause bumps and jolts – can alter the route that the aircraft takes and the feeling of stability for those on-board'. Similarly, seafarers depend strongly on ship technology, while their daily activities on or beneath deck are influenced by the kinaesthetic effects of the ship moving through ocean spaces, as Maria Borovnik explains in Chapter 7. In snowy realms, tyre chains are readily available to make vehicular movement feasible, while in places less used to snow, traffic comes to a standstill. And yet, weather can be seen from different perspectives: snow can also facilitate desired, stimulating mobilities for tourists, as Martin Trandberg Jensen and Szilvia Gyimóthy reveal in Chapter 8.

In moving through the world, weather also conveys other material elements. Wind brings dust and sand that it has picked up elsewhere, perhaps acting as an abrasive force that erodes rocks or stings bodies. Similarly, the movement of weather is impeded, diverted or channelled by materials and forms in the landscape that might divert the direction of wind, sunlight and rain. Air rises as it meets mountain ranges and causes precipitation, rain accumulates and forms streams that pour through valleys and wind howls across plains unimpeded by obstacles. As Tonya Rooney demonstrates in Chapter 4, weather impacts upon rock formations, which may seem to be still, but never are. For rocks are 'ancient and ongoing', very slowly continuing to shift and change shape in response to heat and cold, moistness or aridity, over epochs. Rooney notices that when walking across a rocky landscape, 'water, plants, silty and sandy earth, lichen, creek and bird life seem to grow, die, swirl and move with rocks; forcing, forging, lodging, dislodging, weathering together'.

Weather forecasts alert us to transcontinental movements of weather systems, as animated symbols of warm fronts, depressions and rain belts are displayed on televised maps at national and continental scales. Weather systems shift across oceans and land masses, they defy the notion of any sense of ownership over weather. In recent years, heightened global communication has prompted awareness of El Niño, the vast weather system that emerges in the equatorial Pacific Ocean and generates massive movements of air that shape rainfall levels, cyclone activity and temperatures across large areas of the world. In Chapter 16, Becky Alexis-Martin recounts the turning point in which the human activity of nuclear testing brought about an atmospheric alteration that will exist long beyond our species' demise. She contends that the 'invisible' nature of nuclear warfare masks the terrible mobile impacts of 'nuclear' weather on large populations around the world. Taking a further leap, in Chapter 17, Kimberley Peters reveals that weather does not merely move within Earth's atmosphere but equally swirls around planets of the solar system. And most obviously, the sun is the ultimate cause of all these planetary weather effects, radiating its light and warmth across space, as well as generating the gigantically scaled solar winds that move across the cosmos at extraordinary speeds. Protected from these enormous flows of energy by Earth's atmosphere, particles carried by the solar wind are deflected by the planet's magnetic field, causing the spectacular aurora that plays in the skies above the polar regions. Yet, despite these huge scales at which weather flows, at an immediate sensory level its movement is all too apparent, as blizzards swirl around us, driving rain and sleet force us to turn away from their direction, clouds move across the sky and cover the sun, causing us to don warmer clothing, and wind buffets us as we make our way through space.

Weathered materialities

The rights to land, to hold ownership over the terrestrial, is based on assumptions and a position of the nation-state that 'the earth is a stable platform' that does not move (Bremner, 2020: 10; Steinberg and Peters, 2019). And yet the materiality of Earth's surface is always in motion and weather events such as earthquakes, tsunamis and shifting ocean currents move the earth in ways that make even the largest continental landforms 'drift' (Barry and Keane, 2020: 72–77). Everywhere, weather assails the multiple materialities of the landscape, being absorbed by some elements and absorbing others, penetrating and eroding surfaces and things, and being deflected and repudiated by certain surfaces. As Tim Ingold (2011: 119) insists, the land is not 'an interface' separating earth and sky but is a 'vaguely defined zone of admixture and intermingling'. Weather constantly makes and remakes the material features of a place; wind, for instance, 'scatters seeds, erodes surface material, and shapes the growth of vegetation' (Veale et al., 2014: 26). And the particular material elements that prevail in any landscape respond in distinctive ways to the play of weather. Trees bend in the wind or else they topple, are able to cast off heavy snowfalls or collapse and become buried under snow. And, as we are now witnessing around the world, coastlines are being rapidly eroded by lashing waves and rising sea levels. In her wide-ranging research on monsoon weather in Southeast Asia, Lindsay Bremner highlights how 'processes of weathering, erosion, saltation and alluviation... characterize the dynamic reworking of the earth's surface by weather, wind and water' (2020: 2). Certain kinds of stone and soil can absorb copious quantities of rainfall, or else the earth is fractured in such a way as to allow rain to percolate underground or disperse in numerous streams. Conversely, an incapacity to do so by impermeable or non-porous surfaces such as clay and certain kinds of rock may result in extensive flooding across large areas of land, or in deserts, the creation of deep wadis forged by gushing, highly erosive flows of water. In responding to bright sunlight, snowy surfaces blind, watery surfaces glitter, while forests absorb light and heat. Particular landscapes are thus habitually known, affectively experienced and sensorially apprehended through material qualities that are conditioned by weather and, in turn, influence the range of human activities that can occur.

It is not merely rural landscapes in which materiality is integral to the experience of weather and its effects upon space, for these impacts are also profoundly shaped by the materialities and structures of built environments. Urban heat islands are produced through the vast quantity of asphalt, concrete, bricks and tiles that constitute the urban fabric, materialities that cannot absorb heat like vegetation when the sun beats down on the city. Large towers, when clad with glass can divert sunlight in unexpected ways, such as the so-called walkie-talkie skyscraper in London, built in 2014, that focused reflected sunlight onto cars in the streets below the building, causing them to melt and buckle from the heat. Less dramatically, the clustering of large tower blocks can perennially cast certain parts of the city in shadow and their configuration may generate powerful, unsettling wind tunnels. The impacts of inaptly positioned objects and unsuitable material usage draw attention to how particular substances have been historically deployed in places to minimize the harsh effects of weather. For example, North African casbahs possess extremely thick walls and small windows that not only offer defensive security but afford cool interior weather in hot desert conditions. In other tropical realms, mud and sand walls, courtyards, shutters, verandas, wind towers and awnings that promote cool air circulation have moderated hot weather in the absence of air conditioning. These examples foretell how building design will have to adapt to the more extreme weather that is emerging through climate change, to deal with increased downpours, droughts, storms, and hotter and colder conditions.

A focus on the relationship between architectural features and weather foregrounds how particular materialities become distinctively weathered. Persistent damp causes brick and cement to crumble, baking heat melts asphalt and flooding washes insubstantial structures away. In Chapter 13, Beth Cullen traces how bricks are a key component of Bangladesh's built environment, and follows the life cycle of a *Bangala* brick from geological extraction to dissolution. Sediment is transformed into clay by annual monsoons, and subsequently mined by over a million seasonal migrant workers before being laboriously transformed into bricks for building, construction and roads. The dust from the bricks suffuses the air, and coats vegetation, fabricated surfaces and lungs, while monsoon rains dampen dust, cleanse surfaces and freshen atmospheres. The ongoing transformation of *Bangla* bricks, and their dependence on seasonal conditions and human action, eloquently highlights how manufactured materials become entangled with weather.

Forms of weather evidently disclose the dynamic vitality of the world, yet in their responses to the effects of weather, it becomes clear that buildings and other structures are also vital; they are aggregations of heterogeneous forms of 'vibrant matter' (Bennett, 2010) that cannot maintain an enduring composition and form forever. The material world is continuously under threat from forces that render it subject to decay and erasure, and primary amongst these forces is weather (Edensor, 2016). Yet weather's impact depends both on the material properties of buildings and their components and on the kinds of weather that assail them. At the famous Angkor Wat temple complex in Cambodia, ruination proceeds swiftly because plentiful rain and humidity provide rich conditions for bacteria, plants and other agents to work on the crumbling stone. In contrast, in Mediterranean climates, Roman concrete structures have lasted for millennia because of the general dryness and warmth.

Sarah Whatmore (2013) reminds us that learning to live with the material impacts of weathering has been firmly implanted in building codes, models and technologies that attempt to predict and mitigate severe weather. But the materialities imbued in less dramatic weather – the ongoing erosion of precipitation, tidal denudation or atmospheric pollution – is not so easy to recognize. For instance, the industrial fumes and coal smoke produced by Manchester's rapid industrial development contributed to the erasure of much of the city's stone fabric (Edensor, 2011). In perpetually altering the material fabric of the built environment, the effects of weather are combated by the critical practices of maintenance and repair, procedures integral to the sustenance of the environments. Following icy winters, roads must be resurfaced; debris needs to be cleared away after the Mauritian cyclones, and high winds in other places remove slates from roofs, which subsequently require replacement; mortar in brick walls has to be repointed after severe effects of ice during cold winters; and house interiors must be disinfected, dried out and replastered after flooding. In a specific example, Gillon and Gibbs (2019) detail how the erosive agencies of salt and abrasive sand-laden winds at new coastal housing complexes in New South Wales, Australia, inform which building materials are chosen, determine their rate of decay and shape the relentless, restorative practices of repair and maintenance that take place when the weather is benign. The maintenance of built structures following weather events are supplemented by the installation of material forms situated to manage the effects of weather. Windbreaks of trees, high walls and hedges are placed in areas of flat land to lessen the impact of wind on crops and buildings; flood basins seek to divert rising waters into low-lying areas to prevent flooding elsewhere; groynes lessen the impacts of high tides whipped by winds to prevent coastal erosion; snow poles identify where engulfed roads run, while gritters and snow ploughs endeavour to keep roads viable in snowy weather.

However, in prioritizing the human requirements of shelter from the weather, such preventative measures often disregard broader ecological impacts. A prime example are the sea walls erected to prevent erosion along Chinese, South Korean and Japanese coasts which have decimated the tidal mudflats and river systems upon which seasonal migratory shorebirds rely. Alarmingly, a sea wall erected in South Korea to avert increasing floods, tsunamis and extremely high tides is estimated to have wiped out almost 300,000 migratory shorebirds (Moores et al., 2016). Elsewhere, the careless removal of infrastructural elements can create future problems. The water tanks or *eris* of Chennai that contained excess water have been infilled and destroyed through urban expansion, practices that Beth Cullen (2019) contends will lead to future flooding. On the other hand, formerly adequate defences that protected against significant weather events may become ineffective with climate change, as was the case with the devastating inundation of sea water that destroyed large areas of New Orleans following the advent of Hurricane Katrina. These examples highlight Whatmore's important contention that knowledge of the environment – and particularly of weather disturbances – requires a relational understanding of materiality that goes beyond the human (2013), an assertion that is more urgent than ever.

Affective engagements with weather

Weather not only activates us to think of adjusting clothing or deciding whether to move outside or not; as Ingold and Kurttila (2000: 187) insist, weather is defined by 'what it feels like to be warm, or cold, drenched in rain, caught in a storm…', sensory and affective entanglements that foreground the more-than-representational ways in which we live with weather. Weather affects our moods and emotions – as Luke Howard observes in 1802, noting that clouds 'are commonly […] good visible indications of these causes (of rain and other weather) as is the countenance of the state of a person's mind or body' (in Barnett, 2015: 77). Howard further considers that 'clouds are constantly changing, merging, rising, falling and spreading throughout the atmosphere, rarely maintaining the same shapes for more than a few minutes at a time' (in Hamblyn, 2017: 49). Hamblyn regards weather as 'connected stages in a dynamic process' that is part of other interconnections that include human moods. In 1822, Goethe similarly draws comparisons between cloud formations and moods in his poem, 'In Honour of Howard' (Howard's Ehrengedächtnis), writing that stratus cloud formations cause us to 'feel, in moments pure and bright as this, the joy of innocence, the thrill of bliss', while when the sky piles up a cumulus-induced mood, 'All the soul's secret thoughts it seems to move, beneath it trembles, while it frowns above' (in Hamblyn, 2017: 53).

A sense of the entanglement of weather with both emotional and corporeal states is provided by common phrases: 'he had a face like thunder'; 'she was full of hot air'; 'he suffered from excessive wind'; 'her tears fell like rain'; 'she has a sunny disposition'; 'he was cold and uncaring'. Familiar reference points are also part of an affective and sensory belonging to place. As Lucy Lippard claims, if 'one has been raised in a place, its textures and sensations, its smells and sounds, are recalled as they felt to child's, adolescent's, adult's body' (1997: 34). Deepened by time, such familiar weather-worlds are embedded in sensory and affective memories. These familiarities are captured nicely by Marie Madzak (2020: 197) in her examination of holidaymakers at a Danish caravan site. She characterizes the weather in Denmark as 'ever changing from blue to grey, from cold and crisp to warm and wet; it is ever present, always surrounding us, permeating our walls and clothes and affecting our moods and doings'. This multisensuous, reciprocal interplay between caravanners and weather includes a revelling in the 'the reflections of the colours of the sky in the water, and the dense and wet, dark sand' on the beach (ibid: 205) and luxuriating in the sounds and smells of the rain, while simultaneously deploying heaters and socks to facilitate the production of a cozy (or hygge) atmosphere in the caravan's dry interior. The affective experience of weather can also be accessed by moving to unfamiliar realms as explored in Martin Jensen and Szilvia Gyimóthy's study in Chapter 8, which reveals that Indian tourists visiting Swiss ski slopes playfully move through and manipulate the unfamiliar material affordances of snow. Recalibrating to unfamiliar weather, it is the delightful, sensuous snow itself that invites Indians to this process of becoming competent snow tourists and learning how to orient themselves through a snowy landscape.

The experience of these non-representational affordances is appositely captured by Derek McCormack's (2009) conception of atmosphere that enfolds both meteorological and affective registers, in which a 'geopoetics of air' describes how air is apprehended as a 'constitutive and turbulent participant in the distributed natures of lively worlds' (2009: 39). This foregrounds the capacity of air – as well as other forms of weather – to blur distinctions between bodies and environment, between inside and outside, wherein air is both charged with a sensory intensity that is discerned as part of a dynamic atmosphere, and also possesses particular qualities of moisture, thickness and movement that is physically entangled with an affective ambience. An 'airy poetics' is further elaborated upon by Sasha Engelmann (2015: 432) in her depiction of how air generates affect that 'condenses, environs, ventilates and dissipates around human and nonhuman bodies and objects'.

Such entanglements with weather induce us to become sensorially and affectively attuned to place and trigger particular responses when familiar feelings are solicited by cold, warmth, light, wind, colour and dampness. Yet a reliance on knowing and feeling familiar space, our comfort zone, is increasingly being disrupted by the odd weather that sporadically arrives to thwart expectations. Normative ways of inhabiting place must incorporate the unanticipated advent of storms, floods, bushfires, downpours, tornadoes, droughts, heavy snowfalls and heat waves that are increasing as a result of human-induced climate change. As we write, there has been a radical transformation in the weather as a consequence of the drastic reduction in vehicular traffic, industrial production and aeroplane flights during the lockdown imposed by many governments during the COVID-19 pandemic crisis. As a consequence, weather sensations in cities such as London or Beijing have drastically changed. Thick air pollution that typically saturates the skies with particulate matter, making breathing difficult and visions of the city from afar murky, has been replaced by a lucid sky as carbon emissions plummet, also reducing the risks of asthma, heart attacks and lung disease. Reports from the city of Jalandhar in Punjab, India, focus on inhabitants' astonishment at views of the Himalaya mountains that have been imperceptible for decades due to thick air pollution, now absent with industrial shutdowns and restricted transport availability.

In the face of such adverse and unpredictable weather, Eliza De Vet (2017) shows that people retain a desire to be 'weather connected', to stay positively acclimatized. Indeed, weather possesses an imaginative and affective power despite the dramatic effects of climate change, so that humans continue to situate themselves in place as 'weatherculturalists' (Hulme, 2017). Nonetheless, as de Vet (2017) contends in her discussion of adaptive responses by local inhabitants to changing weather in Darwin, Australia, they recognize that it will invariably be challenging and consequently develop resilience and preparedness. This is especially so among farmers, rural dwellers, fisherfolk and seafarers, who cannot remain insulated against weather. For these people, preparedness becomes part of the shifting qualities of place, and intensifies place 'attachment' through sensory immersion. Amongst seafarers, an affective weather and place attachment intersection is complex, as Maria Borovnik explores in Chapter 7, as they shift between changeable ship environments, ports, oceans and homes across the seas. The interplay between ocean and weather needs ongoing adjustments and is continuously apprehended visually, audibly and kinaesthetically, as the ship moves and responds to changes in temperature and movement (Borovnik, 2017). This compendium of experiences, spanning changes and extreme weather over time has solicited an intuitive ability to read and anticipate weather amongst seafarers, that include the interconnections between sound and kinaesthesia that emerge on the ship during a hurricane (Borovnik, 2019). However, the interplay of weather elements can also be a playfully disruptive experience, as explored in Chapter 7, where ships stuck in ice are momentarily on hold, offering possibilities for fun instead of the usual grinding work of everyday seafaring.

Another example of the entanglement of people with the affective force of weather is observed in frequent cyclones which hit the island states of Mauritius, Haiti, Samoa and Vanuatu, sometimes with devastating consequences, as with the catastrophic events of the 2004 Aceh tsunami disaster in the Indian Ocean. Early warning, careful preparations for 'torrential rain, high winds and flooding' ('Samoa prepares for Cyclone Heta', 2004), and local knowledge of the changes in ocean and weather are critical and may be life-saving. The following example shows how such extreme weather can foster a familiar sense of affective and sensory togetherness.

In Mauritius, cyclones are fairly frequent between January and March. These storms bring high waves, powerful winds and abundant rain, and pose a threat to infrastructures, vegetation and homes. Before the cyclone arrives, the airport is closed and flights cancelled, while schools, shops and businesses are closed. Mauritians become habituated to responding to warnings of coming cyclones on television and radio. By telephoning a free number that advises about the storm's progress, they follow a series of stages to secure their safety and ensure that they have a portable radio and fresh batteries, torches and candles, fresh water, basic and tinned food, can openers, a stove with sufficient gas and a first aid kit and tool kit for emergencies. The cyclone passes through four stages of intensity and threat. When the cyclone reaches Class 1, it is advisable to secure doors and windows with shutters, and all boats moored offshore are brought inland. When the event has attained Class 2, more stringent procedures are called for, with all loose items around property secured and tree branches trimmed. Rural inhabitants that live in less-stable dwellings may search for shelter in public buildings or churches. At Class 3, domestic animals are sheltered and cars are prohibited from roads. It is inadvisable to go outside. When the cyclone reaches Class 4, possibly with gusts of wind up to 120 km, residents seek shelter in the safest part of the house, disconnect all electrical appliances, and if in danger from property damage, protect themselves with mattresses or blankets, while continually keeping abreast of weather bulletins.

Although the advent of a cyclone may be alarming, it is also exciting and causes an animated ambience amongst many households, spreading a shared structure of feeling. The usual routine of the day has been transformed, and anticipation of the storm is stoked by the gathering clouds offshore, progressively heavy rain and increasingly powerful winds. The decision to enter the house is accompanied by foreboding and a sense of thrill is solicited by the cacophonous sound of wind and rain lashing buildings and the surrounding land. The event often takes place in darkness or candlelight, further intensifying the domestic atmosphere. The morning after, people expectantly step out of their house to survey the damage, often greeting a scene in which objects, sand and vegetation lie scattered across familiar space, making it strange. In short, the period including the build-up, advent and aftermath of the cyclone constitutes a profoundly sensory and affective shared occasion, a crystallizing of the potent entanglements of weather with everyday lives.

Representing weather

Finally, we want to emphasize that one way in which weather is thoroughly embedded in place is through the protean and diverse ways in which it serves as a subject for representation across both high and popular cultures. In this book, chapters explore diverse spatial settings in which weather constitutes an integral part of how they are construed, represented and experienced. These encompass different national, regional, rural and urban circumstances, along with physical geographical features, together with 'embedded cultural knowledges, norms, values, practices and infrastructures' that shape shared and individual responses to weather (Veale et al., 2014: 26). Such depictions inform how weather is perceived and sensed and underpin broader cultural assumptions about the qualities of place.

Tim Edensor (Chapter 10) discusses how the late 19th-century Heidelberg School painters endeavoured to represent the vibrant light of Victorian summers, moving away from prevailing practices that portrayed these landscapes according to more muted European aesthetics. In so doing, they more accurately captured a distinctive Australian variant of seasonal radiance. Painterly approaches to weather are famously exemplified in John Constable's sustained attempt to identify the manifold play of clouds and light across southern England, with a particular focus on the skies of Salisbury, Suffolk, Hampshire and Hampstead Heath, with the latter represented in more than 100 paintings created at different times of day and season. Constable's aesthetic concerns were allied to his acute scientific interest in the dynamic meteorological processes that produced such effects and were termed by Kenneth Clark as 'the romantic conjunction of science and ecstasy' (cited in Thornes, 2008: 572). Pyrs Gruffudd (1991: 19) cites the author of a 1946 book on weather who proudly proclaims, 'it is this country with its changing skies and flying cloud shadows that has produced Wordsworth, Constable and Turner'.

A more contemporary account of representing weather in painting is provided by Susan Conway's (2010: 241) description of attempting to paint the tropical feel of Thailand's Chiang Mai Valley:

> When the rains come, the sky becomes a threatening indigo with intervening flashes of lightning. The clouds are so heavy you can feel them pressing down like weights on the land... You get this fantastic contrast, indigo sky, indigo water, and shards of light, among the parallel rows of tiny acid-green plants. When the sun breaks through, the equilibrium between sky and land seems to be restored. Light breezes ripple the surface of the water, tropical insects skim across it, and fish create small swells. The clouds reflected in the water appear softer and lighter.

She continues by explaining how she sought to represent these dynamic scenes:

> Layers of stained canvas were overlaid with dyed silk. This had to be worked until the balance of the underlying canvas (representing the sky) balanced with the overlaid silk (the water in the fields). A third layer was the painting of the rice plants: small and brilliant in the early season, then tall, straight and a duller green, and finally gold as they ripened.

Other creative expressions seek to represent weather that goes beyond classical artistic expressions. The work of artist Roni Horn has involved an ongoing fascination with weather, assembled through oral histories and photographic documentation. Horn (2007: 9) reminds us of the communality that weather brings:

> Everyone has a story about the weather. This may be one of the only things each of us holds in common... Small talk everywhere occasions the popular distribution of the weather. Some say talking about the weather is talking about oneself.

Importantly, Horn (ibid: 10) notes that, 'Weather that is nice is often weather that is wrong. The nice is occurring in the immediate and individual and the wrong is occurring system-wide'. This comment is echoed in Horn's artworks on the swirling depths of oceans and water, the transformative cycle of water through weather systems, and across geological ages as it is preserved in ice, revealing the often-limited social understandings of weather beyond human needs. While a fleeting moment is captured in an impressionist painting of sunshine breaking through clouds, or we bathe in sunshine while exercising outdoors, these individual experiences are often disconnected from the larger system-wide conditions. Here, issues of representation and how 'nice' weather – sunny, clear skies with moderate temperatures, the kind that weather reporters salivate over as optimal for our weekend leisure – is often a sign, in many parts of the world, of a lack of rainfall, prolonged drought and disruptions to expected weather and seasonal patterns.

Creative projects draw attention to the intensities of weather and the affects it has on our bodies. In Chapter 11, Maria Borovnik and Kaya Barry outline some of the poetry, fiction and artistic projects that have engaged with foggy weather, as well as large-scale art projects that conjure 'artificial' forms of weather for public audiences, conveying sensory moods. Exemplary here are many projects by the artist Olafur Eliasson, who has recreated mists and rivers, and harvested blocks of ice and transported them across the seas, in producing a series of immersive, highly sensual and large-scale art installations. These artworks draw attention to how weather, in its un-engineered occurrences, often eludes our attention. Yet spectacular weather surrounds us – not necessarily in melting ice sheets – but in a sunrise after a stormy night. Yet, Eliasson's artistic works complicate our relationships to weather, as with his *Ice Watch*, a work heavily critiqued for the significant materials required as well as the emissions produced by shipping 110 tonnes of ice from Greenland to Paris and London. In this way, Barry and Keane (2020: 201) note:

> the artwork holds up *nature* as something to be admired for the rarity and beauty, separated out from an individual human's actions as a spectacle. After all, simply being a bystander of this artwork and watching the ice melt does not cause the ice to melt; nevertheless, the viewer is made complicit through the act of observation.

It is precisely these 'taken for granted' relationships to weather that creative representations capture so well, perhaps most explicitly in Eliasson's *Weather Project*, a huge installation commissioned for London's Tate Modern gallery, which was characterized by the reproduction of a huge sun at one end of the hall, often bathed in a fine haze to create a glowing, immersive realm in which visitors were mesmerized, often lying on the floor. The artificial sun, skillfully rendered by an assembly of materials including lights, mirrors, projectors and a haze machine, attracted huge audiences to witness the audacious attempt to recreate that which cannot be recreated: the sun. In a similar vein, in Chapter 6, Kaya Barry discusses an artwork she created that captures the sensations of turbulence during air travel, redistributing sensory information that cuts across human bodies, materials and the weather that we are always moving through. Recording the vibrations and resonances during a flight in a severe tropical low in Australia, the weather data that Barry's artwork (re)presents offers a way to consider connections to weather systems that go beyond human sensory capacities.

Many filmmakers are also concerned with depicting dramatic forms of weather to enchant the mood of the movie, using weather events to drive narrative and set ambience and mood. Notable examples are Gene Kelly's renowned dance scene in *Singing in the Rain,* the torrential rain washing away the filth that covers the protagonist after his escape from a sewer in *The Shawshank Redemption,* or the recent highly acclaimed *Parasite* by Bong Joon-ho in which the monsoonal rains flush out a lower-class family from their sub-basement apartment but is conceived as a blessing by a rich family. But such scenes are not simply passively consumed by viewers. In Chapter 8, Martin Jensen and Szilvia Gyimóthy highlight how weather in Asian films produce popular cultural representations and romantic imaginaries that inspire Indian tourists to photographically record their playful posing in imitation of renowned romantic scenes from Bollywood films set in the Swiss Alps. These ludic reenactments portray the snowy landscape backdrops in stark contrast to Eurocentric representations of the Alps as spaces of heroic adventure and conquest.

Literary accounts that deploy weather to enhance the mood, narrative, character and location in which stories are set are also numerous. Three examples are the wild, stormy scenes set on the moor in Emily Bronte's *Wuthering Heights*; the softly falling snow in James Joyce's short story, *The Dead*; and the dense tropical heat that suffuses Joseph Conrad's *Heart of Darkness*. In Chapter 9, Gareth Hoskins details the numerous novels, short stories and films that have used the Santa Ana wind as a metaphorical trope to enliven and contextualize stories of passion, criminality and madness. In addition to these fictional representations, the weather frequently looms large in popular music, with abundant references to rain, sun, wind and heat in songs, as with the aforementioned 'Singing in the Rain', referenced in Chapter 4 by Susannah Clement as a young child expresses her pleasure in spending time in a downpour. This exemplifies how cultural and creative representations intermingle with understandings, practices and affective engagements with weather.

Reconsidering the spatial contexts of weather

In this book, chapters explore spatial, mobile and affective ways in which weather is construed, represented and experienced. As we have emphasized, although it never accords with bounded spaces, weather is bestowed with cultural, social and political resonance by its incorporation into national, regional and local spatial entities – scalar epistemologies through which we make sense of our everyday experiences of weather. As Michael Billig (1995) contends, national meteorological infrastructures configure weather into a national context in accord with conventional modern constructions of belonging. Smaller spatial realms are also assigned distinctive weather characteristics. In Suva, Fiji, like clockwork at 5pm every day, a torrent of rain comes down during the hot season; Wellington, New Zealand and Chicago,

USA, are both labelled the 'windy city'; while Manchester, England and Seattle, USA are renowned as consistently rainy cities. The valley of the Mohaka River in New Zealand is named Te Hāroto, 'the lake of breath', because fog regularly sits upon it, whereas Melbourne, Australia, is reputed for its rapid changes in temperature – inspiring the cliché that one can experience four seasons in a single day. Certain weather phenomena become associated with particular areas, as with diverse winds. The Sirocco originates in the Sahara Desert and brings cool, wet weather to southern Europe, while the Mistral commences in the northeast and powerfully moves through the Rhone Valley to bring cold air to southern France. As Veale et al. (2014) describe, the more localized Helm Wind blows down the southwest slope of the Cross Fell in Cumbria. In Chapter 9, Gareth Hoskins eloquently details the powerful Santa Ana wind that blows from the desert conditions of the Great Basin into the coastal areas of Southern California, and generates potent myths, fictional narratives and scientific studies that explore its impact on human behaviour. As Sarah Strauss (2007: S166) asserts, 'often, these winds have names, testaments to the strength of their identities, each recognizable from afar by internal or external indicators, such as a "feeling in the bones" or a particular pattern of cloud cover'. They are anticipated by inhabitants and contribute to place myths. Moreover, an attunement to distinctive local weather qualities solicits specific words or phrases. For instance, while a powerful storm is termed a hurricane in regions adjacent to the Atlantic and the Northeast Pacific, similar forces are named cyclones in the South Pacific and around the Indian Ocean, or typhoons in the Northwest Pacific (Maloney, 2017). Emphatically, weather remains thoroughly tethered to local, regional and, above all, national geographies.

Yet, by going well beyond the national, in Chapter 17, Kimberley Peters explains that weather may be explored in deep space and time. In taking an outer space perspective to consider how weather moves, she asserts that we can explore new ways in which to comprehend the gigantic scales at which weather emerges and dissipates, and argues that Ingold's notion of weather's entanglement within an open world can be extended to incorporate an extra-planetary openness. The melding and interpenetration of earth and sky, she contends, extends beyond the earth into the extra-planetary realm. Everyday weather cycles are influenced by the earth's movements in space, and atmospheric conditions are shaped by the sun's radiation. Accordingly, although citizens tend to refer to the weather as part of a national homeland in which they pursue everyday habits and routines, this extra-planetary view suggests that weather unites us as earth-beings. In Chapter 4, Tonya Rooney points to the ancient, slow, epochal movements of rocks, noting that these and children's drawings of storms represent lively connections between the present and an unfathomable past. Although the reporting of weather is taken from satellite information, circling around the globe, it continues to be presented nationally or is described and circumscribed in terms of its regional or local significance.

In this collection we have sought to transcend these circumscriptions by emphasizing that contemporaneously, the experience and meaning of weather is saturated by highly diverse spatial contexts, cultures and scales. In an Australian context, there are small towns ravaged by fire, Indigenous connections to country that resist postcolonial narratives of weather, provincial towns with frequent rainfall, diverse areas of the continent in which particular effects of sunlight reign, children's imaginings of stormy weather, and even airspace that connects distant cities and is one of the world's busiest flight routes. In this book we read of rural farmlands in New Zealand, Danish training routes followed by runners, the global crisscrossing of maritime realms by seafarers who originate in the Islands of the Pacific, the weather effects following nuclear testing, the monsoons that mobilize an immense migrant workforce in Bangladesh, Indian tourists in Swiss ski resorts and winds in Southern California, besides journeys beyond Earth into planetary realms.

References

AccuWeather (2015, February 26). *AccuWeather exceeds 9.5 billion requests for global data every day, setting new record in big data demand.* Available at: https://www.accuweather.com/en/press/43009943

Badlan, R (2019, November 12). Firestorms and flaming tornadoes: how bushfires create their own ferocious weather systems. *The Conversation.* Available at: https://theconversation.com/firestorms-and-flaming-tornadoes-how-bushfires-create-their-own-ferocious-weather-systems-126832

Barnett, C (2015). *Rain. A Natural and Cultural History.* New York: Broadway Books.

Barry, K and Keane, J (2020). *Creative Measures of the Anthropocene: Art, Mobilities, and Participatory Geographies.* Singapore: Palgrave.

Bennett, J (2010). *Vibrant Matter: A Political Ecology of Things.* Durham, NC: Duke University Press.

Billig, M (1995). *Banal Nationalism.* London: Sage

Borovnik, M (2017). Night-time navigating. Moving a container ship through darkness. *Transfers: Interdisciplinary Journal of Mobility Studies*, 7(3): 38–55.

Borovnik, M (2019). Endless, sleepless, floating journeys: the sea as workplace. In M Brown and K Peters (eds.), *Living with the Sea: Knowledge, Awareness and Action* (pp. 131–146), Abingdon: Taylor and Frances/Routlege.

Bremner, L (2020). Sedimentary logics and the Rohingya refugee camps in Bangladesh. *Political Geography*, 77: 1–12.

Clifford, K and Travis, W (2018). Knowing climate as a social-ecological-atmospheric construct. *Global Environmental Change*, 49: 1–9.

Conway, S (2010). Painting the weather in the tropics. *Weather*, 65(9): 240–241.

Crouch, D (1999). Introduction: encounters in leisure/tourism. In D Crouch (ed.), *Leisure/Tourism Geographies: Practices and Geographical Knowledge,* London: Routledge, pp. 1–16.

Cullen, B (2019). Haunted landscapes: ghosts of Chennai past, present and future yet-to-come. In L Bremner (ed.), *Monsoon [+ other] Waters.* London: University of Westminster, pp. 185–200.

de Vet, E (2017). Experiencing and responding to everyday weather in Darwin, Australia: the important role of tolerance. *Weather, Climate, and Society*, 9(2): 141–154.

Edensor, T (2011). Entangled agencies, material networks and repair in a building assemblage: the mutable stone of St Ann's Church, Manchester. *Transactions of the Institute of British Geographers*, 36(2): 238–252.

Edensor, T (2016). Incipient ruination and the precarity of buildings: materiality, non-human and human agents, and maintenance and repair. In M Bille and T Sorensen (eds.), *Assembling Archaeology, Atmosphere and the Performance of Building Spaces*. London: Routledge, pp. 366–382.

Engelmann, S (2015). Toward a poetics of air: sequencing and surfacing breath. *Transactions of the Institute of British Geographers*, 40(3): 430–444.

Frykman, J and Löfgren, O (eds.) (1996). *Forces of Habit: Exploring Everyday Culture*. Lund: Lund University Press.

Geoghegan, H and Leyson, C (2012). On climate change and cultural geography: farming on the Lizard Peninsula, Cornwall, UK. *Climatic Change*, 113(1): 55–66.

Gillon, C and Gibbs, L (2019). Coastal homemaking: navigating housing ideals, home realities, and more-than-human processes. *Environment and Planning D: Society and Space*, 37(1): 104–121.

Gruffudd, P (1991). Reach for the sky: the air and English cultural nationalism. *Landscape Research*, 16(2): 19–24.

Hamblyn, R (2017). *Clouds*. London: Reaktion Books.

Hitchings, R (2010). Seasonal climate change and the indoor city worker. *Transactions of the Institute of British Geographers*, 35(2): 282–298.

Horn, R (2007). *Weather Reports You*. Göttingen: Steidl.

Hulme, M (2017: 27). *Weathered: Cultures of Climate*. London: Sage.

Ingold, T (2007). Earth, sky, wind, and weather. *Journal of the Royal Anthropological Institute*. 13: S19–S38.

Ingold, T (2011). *Being Alive: Essays on Movement, Knowledge and Description*. London: Routledge.

Ingold, T and Kurttila, T (2000). Perceiving the environment in Finnish Lapland. *Body and Society*, 6: 183–196.

Ismail, N (2020). Refugees must be protected during the coronavirus pandemic. *Al Jazeera*. Available at: https://www.aljazeera.com/indepth/opinion/refugees-protected-coronavirus-pandemic-200422101438710.html

Levin, S (2020). "If I get it, I die": homeless residents say inhumane shelter conditions will spread coronavirus. *The Guardian*. Available at: https://www.theguardian.com/world/2020/mar/19/if-i-get-it-i-die-homeless-residents-say-inhumane-shelter-conditions-will-spread-coronavirus

Lippard, L (1997). *The Lure of the Local: Senses of Place in a Multicentered Society*. New York: The New Press.

Madzak, M (2020). Weaving through weather on a Danish Caravan Site. *Space and Culture*, 23(2): 195–209.

Maloney, A (2017). *And Now the Weather*. London: BBC.

McCormack, D (2009). Aerostatic spacing: on things becoming lighter than air. *Transactions of the Institute of British Geographers*, 34(1): 25–41.

Moores, N, Rogers, DI, Rogers, K and Hansbro, PM (2016). Reclamation of tidal flats and shorebird declines in Seamangeum and elsewhere in the Republic of Korea. *Emu – Austral Ornithology*, 116(2): 136–146.

Oppermann, E, Strengers, Y, Maller, M, Rickards, L and Brearley, M (2018). Beyond threshold approaches to extreme heat: repositioning adaptation as everyday practice. *Weather, Climate and Society*, 10: 885–898.

Patrick, WSJ (2015). Pulya-ranyi: winds of change. *Cultural Studies Review*, 21(1): 121–31.

Rancière, J (2009). *Aesthetics and Its Discontents*. Cambridge: Polity Press.

Samoa prepares for Cyclone Heta (2004). Samoa prepares for Cyclone Heta. *New Zealand Herald*, January 5, 2004. Available at: https://www.nzherald.co.nz/world/news/article.cfm?c_id=2&objectid=3542193

Seamon, D (1979). *A Geography of the Lifeworld: Movement, Rest and Encounter*. New York: St Martin's Press.

Steinberg, P and Peters, K (2019). Cross-currents and undertows: a response. *Dialogues in Human Geography*, 9(3): 333–338.

Strauss, S (2007). An ill wind: the Foehn in Leukerbad and beyond. *Journal of the Royal Anthropological Institute*, NS: S165–S181.

Thornes, J (2008). Cultural climatology and the representation of sky, atmosphere, weather and climate in selected art works of Constable, Monet and Eliasson. *Geoforum*, 39(2): 570–580.

Vannini, P and Austin, B (2020). Weather and place. In Edensor, T, Kalindides, A and Kothari, U (eds.), *The Routledge Handbook of Place*. London: Routledge.

Veale, L, Endfield, G and Naylor, S (2014). Knowing weather in place: the Helm wind of cross fell. *Journal of Historical Geography*, 45: 25–37.

Whatmore, S (2013). Earthly powers and affective environments: an ontological politics of flood risk. *Theory, Culture & Society*, 30(7/8): 33–50.

World Meteorological Organization (WMO) (2020). What we do: observations. *World Meteorological Organisation*. Available at: https://public.wmo.int/en/our-mandate/what-we-do/observations

Youatt, R (2017). Personhood and the rights of nature: the new subject of contemporary earth politics. *International Political Sociology*, 11(1): 39–54.

2 Research in weather

Notes on climate, seasons, weather and fieldwork mobilities

Phillip Vannini and April Vannini

Social scientific research on weather and mobilities is growing at a fast rate, as Edensor, Barry, and Borovnik discuss in their introduction to this volume. Research *on* weather, however, is still vastly more common than reflections on the role played by weather *in* research. There is an important difference between the two. Research *on* weather treats the weather as a unit of analysis. Research *in* weather reflexively examines the processes and outcomes of a research project as an experience and practice immersed in a specific weatherworld. Not all research is weather-sensitive, of course, but all fieldwork done in the open is inevitably and deeply affected by weather.

Wind, rain, snow, fog, heat, and cold are shared ongoing concerns in both the planning and execution of all fieldwork and all travel for fieldwork. From the selection of where we do research to what time of the year we do it, and from what particular activities we choose to pursue to how we choose to move around in the field, ethnographers make important weather-, climate, and seasonality-based decisions which deeply affect the outcomes of research. Yet, these considerations often remain undiscussed in the literature. In what follows, we hope to stimulate the conversation on doing research *in* weather by recalling on specific meteorological events and personal experiences and by reflecting on their significance. We focus on a type of fieldwork that is particularly affected by climate, seasonality, and weather: mobile, multisite ethnography.

For the last four years the two of us have been involved in a multisite project on natural heritage, wildness, and wilderness conservation. The fieldwork has taken us to twenty UNESCO World Heritage sites in Canada and around the world. Nearly all our data collection occurs outdoors and in mobile circumstances as part of go-along interviews. Whether we snowshoe, paddle, hike, climb, sail, ice-climb, fly, or walk and talk with our interlocutors, our mobile research is deeply impacted by weather, climate, and seasonality, especially because we film all of our go-along interviews. These conditions have made us particularly sensitive to the numerous ways in which meteorological conditions shape fieldwork practices, experiences, and representations.

To be sure, our current research is not focused *on* weather as a subject. But in studying wildness and wild places as we do, our field experiences are inevitably immersed in the weatherworld. Not only is weather a manifestation of wild nature, but its occurrence can clearly intensify or de-intensify how wild a place is perceived to be. Because our understanding of wildness is rooted in phenomenological experiences, and because such experiences can be so fleeting and ephemeral, the choice of *when* we conduct fieldwork is just as important as the selection of the field sites themselves. Making these decisions carefully is even more important by the fact that our fieldwork is slated to result in the production of two documentary films. When operating expensive audio-visual equipment outdoors (without the benefit of large crews), and particularly when filming in natural settings in remote locations, one cannot afford to be cavalier about the role played by seasonality, climate, and everyday weather. In sum, while not focused *on* weather, our research deeply, intensely, and, at times, dramatically takes place *in* weather.

Our underlying assumptions on weather and the mobilities of fieldwork research are based on Tim Ingold's perspective on the inhabitation of the lifeworld. Ingold argues that weather, place, and being are inseparable, but because they are so deeply knotted together we often neglect to be mindful of their entanglement. Unlike atomizing perspectives that treat weather as an independent variable and human behaviour as a dependent variable, Ingold conceptualizes weather as an inescapable medium for our existence in the world (Ingold, 2005, 2007). Following Ingold's phenomenological and relational perspective, we believe that throughout mobile practices of field research we are immersed within concrete events of weather as a condition of our being-in-the-world. The 'weatherworld' (Ingold, 2005, 2007) is a universe of multiple weather processes that shape distinct weather-places. As part of our inhabitation of the weatherworld we continuously *weather* fieldwork sites through a variety of adaptive practices. But let us begin by first re-thinking of research as a type of mobility, following Tim Cresswell's (2010) notion of mobility constellations.

Fieldwork as a mobility constellation

Cresswell (2010: 18) suggests we should think of mobilities as comprised of six constituent parts: motive force, velocity, rhythm, route, experience, and friction. Taking into account these six dimensions allows us to make sense of mobility constellations as 'particular patterns of movement, representations of movement, and ways of practising movement that make sense together'. Along these lines, we argue that fieldwork is a mobility constellation. The idea is not entirely new. Ever since the publication of Clifford's (1997) book *Routes*, ethnographers across the disciplines have been thinking of fieldwork as a type of itinerant practice similar to travel. This is even more so the case when ethnographers are inspired by the growing tradition of mobile methodologies (see Büscher, Urry, and Witchger, 2010). Thinking of field-work as a mobility constellation is a logical step.

Ethnographers, like all travellers, simply get from one place to another – in their case from their home and office to the field and back (and back again). This is not just physical movement, of course. Ethnographers' meaning-making of their travel practices and experiences results in the writ-ten and audio-visual representations that take shape in published articles, books, and films. Their travels give distinct shape to their own professional and personal identities, and their mobilities play an important role in shaping the subjectivity of the people they encounter on their journeys. Their travels take multiple forms: as adventures, as journeys of discovery, as colonial and imperial pursuits, and, more recently, as democratizing, problem-solving, and emancipatory undertakings.

The mobility constellations of fieldwork have their own constituent parts. Here, we particularly focus on the following:

Motive forces (or what drives ethnographers to do fieldwork): education; discovery; curiosity; career-making; passion for learning and knowledge; collaboration with state and private enterprises; consciousness-raising; emancipation; social amelioration.

Velocities (or how quickly and slowly we conduct our fieldwork): relatively fast (short sojourn ethnography and multisite fieldwork); extremely slow (more traditional versions of anthropologists 'setting up camp' overseas).

Rhythms (how often we go to the field): consistent and repeated (when eth-nographers return to a field regularly over time); inconsistent and repeated (in the case of multi-site researchers who 'follow the thing' or study an is-sue across several sites); inconsistent and unrepeated (as in ethnographers travelling to different field sites over time for different fieldwork projects unrelated to previous ones).

Routes (or the places where we choose to do fieldwork and how we get there): direct; indirect (when ethnographers combine fieldwork journeys with conferences, lectures, or personal holidays).

Frictions (or what slows and stops us down): research permits; visas; inter-view appointments that do not follow through; research collaborations that follow apart; grants that are not obtained or renewed; field sites that become unsafe and unsecure to visit.

Experiences (or what it feels like to be on the field): e.g. ranging from the most personal and professionally rewarding, to the most frustrating.

In sum, we can think of fieldwork not only as a methodological practice but also a mobile one that is shaped by the six typical constituent parts of a mobility constellation. And just like all mobility constellations, fieldwork mobilities are deeply coloured by the weatherworlds in which they take place. In what follows, we will use these concepts to guide our reflection on our own experiences of the relation between fieldwork mobilities and weather.

Frictions

The Overland Track is one of Australia's most famous multi-day hikes. The track criss-crosses Cradle Mountain-Lake St Clair National Park, right in the heart of the Tasmanian Wilderness World Heritage Area (TWWHA). Natural World Heritage sites are typically quite large and it is impossible to experience them in their totality. Because of their size and ecological diversity, we have to choose a specific focus or particular issue and zero in on it, keeping the rest of a particular site in context, but somewhat in the background. In selecting an issue to focus on in TWWHA, we determined that the track's cultural significance as a life-changing experience would be an interesting research focus. In fact, the track is considered almost a rite of passage among Aussie outdoors lovers, as well as an increasing number of visitors from overseas.

Doing participant observation on the track – talking with hikers and guides along the way and filming as much as possible of TWWHA en route – had key practical advantages. The track is typically completed in five or six days – an amount of time that left us with another two weeks to conduct interviews off track, elsewhere in Tasmania. Moreover, long enough (82 km) to be meaningful, but not prohibitively exhausting, the track seemed challenging but not dangerous, relatively undeveloped but requiring no bush-whacking, beautiful, diverse, and teeming with wildlife but promising no encounters with threatening species. The Overland Track thus seemed to be everything that our family could wish to tackle as part of our project on wildness and wilderness conservation.

The two authors of this paper are spouses, and while we are researchers, we are parents first. And as parents, we would not dream of leaving our daughter at home while we travel around the world for fieldwork. We know there is so much she can learn from travel and we are happy to make the occasional accommodations to our itineraries to ensure she can travel with us.

To avoid crowding on the track, a reservation system is in place during summertime. Hikers pay an AUS$100 fee, schedule a departure day, and make a promise not to loiter along the way – thus ending their hike on time five or six days later. We selected 18 December 2015 as our preferred day of departure. Planning the trip was rather complicated. For 18th December, early in the morning, we arranged for a shuttle to pick us up from our hotel in Launceston and drop us off at the park's headquarters at Cradle Mountain. Then we booked another shuttle for the 23rd; this one would meet us at the ferry landing on Lake St Clair's shore and deliver us back to Launceston.

In order to get ready for the trek, we decided to start hiking the trails around our island in British Columbia on a regular basis throughout the fall. To get our bodies accustomed to carrying heavy loads we hiked with heavy backpacks. The rainy season had begun on the West Coast and the trails were occasionally muddy after rainfalls; this too, we thought, was a key terrain challenge to get used to.

By the time we landed in Tasmania, the antipodal spring was ready to give way to summer. Launceston's streets were lined with locals enjoying beer outside pubs and restaurants, retail shops sold beach toys and swim-wear, and cafes attracted passersby with smoothies and ice creams. Next to all this – in a profoundly incongruous manner to our northern hemisphere minds – were pictures of Santa and the elves promoting deals on anything from barbecue grills to surfboards.

Two days later, the 18th arrived. After a smooth highway ride, the shuttle van started to climb the steep, narrow, and winding road to Cradle Moun-tain. Soon enough, poor Autumn found herself spilling the beans on the side of the road, not just once but twice. Motion sickness was not a first for her. But on this day, she was taking it extra hard. Tasmania was recording record high temperatures for this time of the year, and the heat combined with the motion sickness had resulted in dehydrating our poor kid. In a matter of days, she had gone from 0 to 30 degrees.

To give her a little time to recover, we decided to loiter a little bit by the park's headquarters. But the longer we lingered, the higher the sun rose in the sky. It was 33 degrees by the time we laced up our boots. It was now 11 o'clock and we could not afford to delay departure any more. Departing the next day, given the strict reservation system in place and the weather forecast calling for another hot day, was not an option. Our permit and reservation clearly specified that we could only leave that day and that we should reach Waterfall Valley, 10.7 km away, and sleep there overnight. Due to the uphill climbs that journey could take up to six hours, arguably more at our unhurried pace.

The hike started off easy. One kilometer in and our boots had yet to touch the soft tussock laying underneath the long, wide, wooden boardwalk, duly covered in metallic mesh to prevent slipping. It was amidst this, eight min-utes after our start, that Autumn spoke up in a trembling voice.

'I don't feel good.'

Her face was pale and sweaty. She seemed to be shaking. We paused our walk. April spoke to her calmly and gave her some water. We re-assured her that she could do this. Phillip relieved her of her backpack. After a few minutes she stood up and smiled softly. We started walking again. A few steps later, with sweat dripping down her forehead, Autumn spoke again 'I'm sorry. I can't do this.'

Cresswell (2010) suggests that mobilities slow down and stop, at times as a result of choice and at times forcefully and out of necessity. Meteorological conditions that create frictions in mobilities include fog, ice, snow, and high winds, but, at times, heat has also been known to cause mass delays and cancellations. In June 2017, for example, Phoenix airport authorities had to cancel numerous flights as temperatures reached 47 degrees Celsius, making it difficult for airplanes to take off in the thin, hot air. In our family's case, the unseasonal Tasmanian heat caused us to stop our hike on the Overland Track and cancel our plan to reach Lake St Clair. So, instead of focusing our research on the backpackers and the tour guides who hike the trek, we were forced to camp for five days at Cradle Mountain and switch our research focus to the countless boardwalks that we found at Cradle Mountain (Vannini and Vannini, 2018). While Autumn regained strength at our campsite, the two of us cancelled scheduled pick-ups, re-arranged accommodations, planned new research activities and practically re-designed from scratch our entire fieldwork at TWWHA. All of this because of weather.

Doing research outdoors, exposed to the elements, requires coping with the weather and moving accordingly with it. In 2012, Phillip and colleagues (Vannini et al., 2012) argued that dwelling in the weatherworld is akin to what we might call *weathering*. Weather is a noun, but it is also a verb. When we weather a place, we become active in place-making. Weathering a place, a weatherworld, is an emergent and context-rich practice and experience that is very personal. Weathering a hike in 33-degree weather, for example, may be no big deal to an adult resident of tropical environment, but it is extremely difficult for a child like Autumn, who lives in a temperate rainforest. In fact, we may sense weather in the present moment, but our sensations are informed by sensory skills informed by past memories and habits. How we weather a weatherworld like Cradle Mountain is therefore something that is 'produced, enacted and perceived in combination with each other, intertwined with emotion, meaning and memory' (Hsu, 2008: 440). How we deal with the mobility frictions in the weatherworld is thus highly contingent, variable, and demanding of personal involvement and reflexivity.

Routes

'Mobility is channelled,' Cresswell (2010: 24) writes, 'it moves along routes and conduits often provided by conduits in space.' Fieldwork is similarly channelled. The channels are determined by infrastructures, timetables, and so on, but also by less-material (and by no means less-constraining) forces that shape where researchers choose to go and when. Mobile ethnographers who 'follow the thing' (see Cook, 2004), for instance, do their fieldwork alongside routes determined by the pathways their 'things' follow along global channels of production, distribution, and consumption. Fieldworkers who engage in comparative research choose research routes that make sense in the context of their research design. Other practitioners of mobile methodologies may simply follow journeys predetermined in advance by all the stops that a train, or bus, or boat makes. Then there are researchers who choose field sites simply on the basis of what attracts them personally to a place. And as much as personal fascination with a particular issue, cause, or culture matters, many fieldwork sites are also chosen on the basis of climatic conditions. Some are drawn to cold places, others to warm ones.

Imagine you were tasked with studying conservation practices at a coral barrier reef. Certainly, you could choose to conduct interviews in the offices of the people tasked with issuing and managing conservation policies, the offices of NGOs, tour companies, and fishing organizations. But you could also elect to visit that barrier reef yourself to meet with those in charge of enforcing conservation, and also those (like fishers) who have to work and live by those policies. If you were filming, you would most likely want to go to the reef, spend some time there, and boat around atolls to explore and see as much as time and funds allow. And if that was the case you would need to plan your journey with climate data in hand. Travelling during the wet season, you would conclude, would make it nearly impossible to reach certain areas, would result in delays once you reach certain atolls and wait for your ride back, and would put your safety and your equipment at risk.

We decided to travel to Belize in the spring of 2017, having determined that spring was the best time of the year for our needs. As part of our itinerary up and down the barrier reef, we made it a point of travelling to the Great Blue Hole, a massive magnet for nature-based tourists from around the world. Day tours by boat and flightseeing airplanes leave from San Pedro and Belize City carrying sightseers, snorkellers, and divers wishing to relive the images created by the famed Jacques Cousteau. Given our interest in conservation we asked the Belize Audubon Society to take us out on their private vessel. In North America, the Audubon Society is almost synonymous with bird protection but around the world its mandate is much larger. In Belize, the Audubon Society is dedicated to raising environmental awareness and co-managing (among other sites) two of the protected areas inscribed within the borders of the World Heritage Barrier Reef Reserve: the Great Blue Hole and nearby idyllic Half Moon Caye. Indeed, they were happy to show us what they do and their knowledgeable executive director was delighted to jump on the boat with us, together with her extended family.

We departed for the Great Blue Hole on a day that turned out to be as glorious as we hoped. Roughly two hours after leaving the Belize City dock in the early afternoon, the Audubon Society speedboat came to a sudden halt. No sign of land was in sight. The winds were calm, the humidity low, and the azure waters surrounding us had suddenly turned darker in colour. That was the clue: we were at the Great Blue Hole (see Figure 2.1).

'Jump up here with the camera,' the boat captain encouraged Phillip, 'you'll get a good view.' As Phillip made his way to the roof, everyone else on the boat, quickly slipped on fins and snorkels and jumped merrily into the refreshing water. By that time of the day all diving vessels had left, and the Great Blue Hole was ours alone. The season we had chosen to travel to Belize, and the day's weather, made for an unforgettable experience for us and our travel companions as the sunshine, warm temperature, and absence of winds allowed us to fully enjoy the landscape and seemed to put adults and children alike in a pleasant mood.

How we make sense of the weather can be understood as a type of somatic work (Vannini et al., 2012). Somatic work is a range of reflexive experiences and activities by which people create, extinguish, maintain,

Figure 2.1 Blue Hole.
Source: Phillip Vannini and April Vannini.

interrupt, cultivate, and/or communicate sensations that are congruent with personal, interpersonal, and/or cultural notions of moral, aesthetic, and/or logical desirability. Choosing to travel to a place that is likely to yield certain bodily sensations – such as a comforting sense of warmth, or a refreshing feel of coolness coming from diving into tropical sea waters – is a type of somatic work. That much is obvious: few people would willfully create conditions for themselves that are likely to generate unpleasant sensory experiences. Cruises to tropical destinations during the dry season are more or less popular than cruises to the same places during the wet months.

What is less obvious is the role that this kind of somatic work plays in the selection of field sites by researchers. As Ingold (2005, 2007) observes, the weather is not so much something we perceive but rather something we perceive in. How much does the climate of the places where we choose to go affect what we know? We need to be mindful, for example, of what we miss out on when we make decisions to travel to a place at a particular time and not others. Returning to the field at different times throughout the year can be a good idea. While doing so is not possible in our case (because of limited time and budget), we do make it a point to travel to different sites during different seasons. For example, in 2017 and then in 2018, we made the decision to travel to the Canadian Rockies (see Figure 2.2) and Waterton Lakes National Park during the winter precisely in order to balance out the number of trips we had made during spring and summer times. As a profession, we need to be mindful of how much of what we know about the world is influenced by where and when we choose to go do fieldwork.

Figure 2.2 Banff Ice climber.
Source: Phillip Vannini and April Vannini.

Velocity

Much of what we know collectively about mobility and social and cultural dynamics, further following the earlier line of questioning, is affected by how fast
we can get to our field sites. The great majority of social science research takes
place in cities, because there, researchers can get to their field sites quickly.

Infrastructure plays an important part in determining velocity, but so
does weather. How far might you be able to go in 25 hours? It all depends,
one would suppose, on how fast you can get to the nearest airport and hop
on the next intercontinental flight – as long as no storms are causing flight
cancellations. For most people, this kind of velocity is a blessing. As academics, we travel the world as far away as our grant funds allow us to go.
We book flights to conferences in interesting cities. We take out students to
mind-expanding field schools. And we can often be back in the classroom
without having missed too much. But there are some places where few researchers venture because they are too far and inconvenient to reach.

The Ogasawara Islands is one such place, since no way of getting there
exists other than by a ship that leaves Tokyo once a week. With a plan to do
fieldwork in the Ogasawara in December and January of 2018 we boarded
the *Ogasawara-Maru* from Tokyo's Takeshiba Port, loaded with our research
gear and a stock of ginger pills and motion-sickness medicine large enough
to put to sleep a small elephant. Modern, functional, neither luxurious nor

particularly pretty, the *Ogasawara-Maru* made that journey once a week, sailing the 1,000 km separating Tokyo's ward of Koto from Chichijima, the largest island in the Ogasawara Archipelago, in 25 hours.

The islands are inscribed on the World Heritage list due to the abundance of unique endemic species living there. Thanks to evolution unfolding largely in isolation from the rest of the world, and in particular thanks to the absence of human residents well until the 1800s, the 30 islands had long been known as the Bonin Islands – the word 'Bonin' loosely translates as 'uninhabited.' They are inhabited now by about 2,000 people and just a few thousand visitors coming and going throughout the year, mainly arriving from the rest of Japan. There was an obvious reason why visitation numbers are so low, of course. Imagine this.

Imagine sitting on the outer deck of a ship, with nowhere to go for 25 hours. Nowhere to hide. Nowhere to find refuge from the rage of the winds and the anger of the sea. Imagine being under siege by swells so vicious that you do not wish to even look at them. Now imagine the pieces inside your brain – the cerebrum, the brainstem, the cerebellum – all dancing with each other inside your skull, each at a different tune of their own choosing. Now douse that sickening syncopated dance with a melange of malefic odours. The vapours rising from the cookers in the canteen. The stench of shoeless feet and vomit wafting away from the washrooms. The stuffiness breeding within an environment whose windows have been sealed shut to prevent spray from getting everything wet inside. And accompany that image with innards revolting inside your stomach, unable to explode simply because they have been stunned by a near-overdose of drugs meant to crush your consciousness for 25 hours. Imagine all that and you will understand why tourist and research mobilities to the Ogasawara are infrequent and why that place has been uninhabited for so long.

But do manage to put up with all that: the rough seas, the length of time it will take you to get there, and then to wait again to wait for another boat to pick you up, and you will be rewarded with one of the most unique mobility rituals that you will encounter. Any time the *Ogsawara-Maru* departs Chichijima en route back to Tokyo, the locals, knowing very well about the tough journey awaiting the departing boat passengers, gather to bid them farewell, as can be viewed at http://tiny.cc/slfncz.

How much of the world's mobilities are we still unaware of, simply because research journeys to the places where they occur are simply too slow and rough for us? How much does the velocity typical of the research sites most of us travel to influence our worldviews? How different would our comprehension of mobility be if we ventured farther afield, putting up with the slowness of movement, more profound inconveniences and rough weather and climate patterns?

Experiences

Ethnographers write based on what they experience in the field, while ethnographic filmmakers do not have the same privilege. All they can express from their field experiences is what their cameras and sound recorders have recorded. If words were said but never recorded, or were recorded but rendered invisible or inaudible by bad weather, it is as if they were never experienced.

The winds of Patagonia are enough to drive people mad. Early European colonizers, one after another, left tales of memories of winds so strong, so persistent, so loud, that they were eventually forced to abandon their outposts and move away to other parts of Argentina or even return home. The winds are especially vicious during the summer when air masses coming from the Pacific are squeezed through the narrow valleys that snake through the Andes. In towns like El Chalten, locals share stories of finding distant neighbours' kayaks in their backyards after they were tossed around the streets during a windy night. Tour operators even rent bikes one way only and can drop you off in spots where you can conveniently return to town riding a tailwind.

El Chalten is not a place where you want to record go-along interviews if you care about sound quality. But if you carry out fieldwork in Los Glaciares National Park, you have little choice. During winter time the winds are calmer, but snow and freezing temperatures virtually shut down the small town of El Chalten. Travel, for research or tourism, becomes nearly impossible. Summer is the only option.

What people do in order to weather places is a form of somatic work. To weather a place means to expose yourself to its elements, to undergo exposure to its challenges, to strive to adapt, and to make sense of an environment by understanding it, making it familiar, sensible, intelligible, and somehow relatively predictable. The weatherworld, borrowing from Ingold, is a taskscape (2000). So, to weather a weatherworld in which you are immersed means to prepare, to struggle, to learn to endure, and to strive to cope with climatic elements. Accordingly, a good deal of what we do in our field research to cope with an unfamiliar meteorological environment can then be understood as a taskscape.

We tried to cope with the Patagonian winds. Because it is extremely challenging to 'clean' wind noise from a sound recording during editing, we did everything we could to protect our microphones from the winds. We bought extra lav microphones that are less sensitive to wind than shotgun mics. We invested in extra fuzzy windscreens intended to keep lav mics safe from the rumble of the wind. We learned to hide those mics under clothing. We reminded ourselves to only ask the important questions when an interviewee's back was facing the wind, with their body acting as a shield for the recording. We learned to move and talk together with the winds' movements, with their own chatter. We used our senses as skills (Grasseni, 2007).

Yet the key point is that there are a few things we would do differently if we returned. How we weather a place the first time we visit it is different to how we weather it on successive occasions, as our experience accumulates. Repeated journeys build familiarity, allow us to generate more accurate expectations, and increase our skills. Our experience or lack thereof with meteorological conditions depending on the frequency of travel to a field site has important consequences for the outcomes of our fieldwork. In our case, while we had never been to Patagonia, our experience with similarly windy sites had prepared us to weather the place and record our interviews effectively.

How much of our collective research is filtered by our ability or inability to weather a site? How much does our travel experience affect how much we can learn, and what we can learn at a field site? How large an effect do meteorological conditions, tempered by previous experiences or lack thereof, have on research journeys' outcomes?

Rhythm

All ethnographers know that their research depends on the goodwill of their interlocutors. The people that agree to make time for us by showing us what they do, leading us to places that matter to them, meeting with us to talk and to teach us whatever we wish to learn from them are individuals who, like the rest of us, lead busy lives. To recognize the value of their knowledge and their time we make appointments with them and do our best to show up punctually and not abuse their kindness by taking more of their time than they agreed to give us. Punctuality and efficiency in scheduling an appropriate number of evenly spaced out meetings with people becomes especially important where fieldwork is bound by a definite, short time period. When you are in a field site – perhaps even a small one – for a year or two, you have greater chances of running into someone repeatedly, of making time to talk whenever time can be spared. But if you travel to many different sites, you need to be careful with your time and your money. Take our recent fieldwork trip to Iceland, which took place in March/April of 2019.

Because the research grant enabling our research was issued in Canadian Dollars, our trip to the land of the Icelandic Krona had to be tight and well-planned. This meant having sit-down interviews and go-alongs planned with maps and driving distances in mind weeks ahead of time. And yet, such planning ahead in a place like Iceland means that you take your chances with the weather. And not just any weather; springtime Icelandic weather. A 30-second clip gives an idea of the weather to which we are referring, as Phillip struggles to walk: https://vimeo.com/325909381

Some Icelanders tell us that much of the country's economic history of sudden booms and busts can be understood in the context of the traditional opportunistic culture of fishing folk. As fishers do when they see suddenly clearing skies, politicians and entrepreneurs learn to do when they see a chance. Then, when the winds change for the worse, they learn to alter course just as quickly. When you get in the habit of changing directions as quickly as the weather shifts, your collective rhythms eventually become habits, worldviews, and cultural customs.

Immersed in the Icelandic weatherworld, we quickly learned to function the same way. We learned to regulate our fieldwork rhythms not so much by the rhythms of our schedule set weeks ahead but by the rhythms of the changing skies. Much to our surprise, our interlocutors patiently understood and accommodated us. Interview times were rebuilt the day before on the basis of forecasts and road conditions. Field outings that began as snowfall bursts stopped were later paused as windstorms erupted, only to resume as the sun peeked through the clouds, and were interrupted again when horizontal rain pelted our cameras and our faces. Like never before, our ethnography assumed a distinct *staccato* rhythm, with any few minutes of good weather and interrupted go-along interviews feeling like a *tempo rubato* – a time stolen – from the skies.

Making sense of weather, even in places like Iceland where the weather seems to make little sense to outsiders, means setting up the rhythms of research and travel to match meteorological rhythms. Inhabiting the weatherworld is a skilled practice that shapes our place in the world (Ingold, 2010). We weather this world reflexively, individually, and collectively. We perceive what the weather does to us and those around us, and learn to co-exist with it, to move alongside with its movements, to take time to learn as it allows us to do so. We recall the weather of the past and compare it to the present, often in order to predict the future. We make sense of it, plan in accordance with it, and reflect on risks and rewards. 'We are our weather,' writes Berland (1993: 207). Its movements are our rhythms, its directions our itineraries, its comings and goings our risks and opportunities.

Motives (and a few concluding thoughts)

They say that a picture is worth a thousand words, and so we refer you to the picture below, which we offer as part of our conclusion (see Figure 2.3). It was taken near Kathleen Lake, Yukon, just outside the boundaries of Kluane National Park. Clustered near a few other national and provincial parks in British Columbia, the Yukon, and Alaska, Kluane National Park is part of the second-largest World Heritage site in the world, an area over twice the size of Switzerland. Because the conditions of our grant stipulated that we had to conduct fieldwork at all the ten UNESCO natural World Heritage sites located in Canada, we never quite 'chose' to travel to Kluane. We 'had to.' But did we really 'have to' in the sense of a compelling imperative?

Like one of the editors of this volume, we have a fascination with light and darkness (Edensor, 2017), and especially with the Northern Lights: the Aurora Borealis (Edensor, 2010). In World Heritage sites like Canada's Wood Buffalo National Park, Aurora Borealis–watching has become such a popular activity that park managers have been seeking ways to preserve the darkness of the night sky, much like one would preserve other forms of wild nature. A similar appreciation of darkness has resulted in Kluane National Park becoming a mecca for many overseas tourists fascinated with the display of light accompanying nighttime solar winds. Indeed, many people choose to travel for the sake of cultivating sensations arising from exposure

Figure 2.3 Northern Lights in the Yukon.
Source: Phillip Vannini and April Vannini.

to specific weather – and season-related – events like the Aurora Borealis or maple leaves changing colours in the fall. But is it as obvious that researchers choose to travel to field sites for the same reasons?

A possible answer is that no, researchers are moved to travel by 'higher' motives, and perhaps they do not quite 'choose' to travel for field, as they are compelled by a greater call. 'Of course,' observes Cresswell (2010: 23), 'the difference between choosing and not choosing is never straightforward and there are clearly degrees of necessity' in what motivates mobilities for fieldwork or otherwise. But while the need to collect and accumulate knowledge for the sake of public education, or the will to solve specific problems and ameliorate social conditions are strong forces in motivating ethnographers to travel to select field sites, weather, climate and seasonality are clearly important, at least in guiding practical motives, as well. We may not always recognize it, but our choice of specific research projects and related sites, what we do there, and how we do it, is clearly affected by meteorological factors.

Field research is a type of travel: a unique mobility constellation. As such, research has its velocities, rhythms, routes, experiences, frictions, and motive forces. And like other mobility constellations, the practices, representations, and experiences of field research are profoundly coloured by the climate, seasonality, and weather conditions in which they take place. As we have argued throughout this chapter, doing research entails weathering a place, making conscious and sometimes not so conscious efforts to cope, enduring and adapting to meteorological and methodological taskscapes. Weathering is a type of somatic work, an ensemble of activities by which we manipulate and cultivate our weather-related sensations. Doing research in the open is simply unthinkable in abstraction from the kind of somatic work we do in weathering our field sites.

Weather, place, and being are inseparable, as Ingold (2010) has convincingly argued. Weather, place, being, and *knowing* are also inseparable. We know as we go but often neglect to be mindful of the many ways in which our entanglements in our lifeworlds matter. Therefore, we call upon mobile researchers, fieldworkers, and ethnographers to become more highly attuned to our immersion within concrete meteorological events. Weather, climate, and seasonality are an inevitable condition of our being-in-the-world, as much as they are of our moving-with-the-world, and researching-, and knowing-*in*-the-world.

References

Berland, J (1993). 'Weathering the north: climate, colonialism, and the mediated body'. In: V Blundell, J Shepherd and I Taylor (eds.), *Relocating cultural studies: developments in theory and research*. Routledge, London, pp. 207–225.

Büscher, M, Urry, J and Witchger, K (eds.) (2010). *Mobile methods*. Routledge, London.

Clifford, J (1997). *Routes*. Harvard University Press, Boston.

Cook, I (2004). Follow the thing: papaya. *Antipode*, 36(4): 642–664.

Cresswell, T (2010). Towards a politics of mobility. *Environment & Planning D*, 28(1): 17–31.

Edensor, T (2010). 'Aurora landscapes: affective atmospheres of light and dark'. In: K Benediktsson and AK Lund (eds.), *Conversations with landscape*. Routledge, London, pp. 227–240.

Edensor, T (2017). *From light to dark*. University of Minnesota Press, Minneapolis.

Grasseni, C (ed.). (2007). *Skilled visions*. Berghann, New York.

Hsu, E (2008). The senses and the social: an introduction. *Ethnos*, 73(4): 433–443.

Ingold, T (2000). *The perception of the environment*. Routledge, London.

Ingold, T (2005). The eye of the storm: visual perception and the weather. *Visual Studies*, 20(1): 97–104.

Ingold, T (2007). Earth, sky, wind, and weather. *Journal of the Royal Anthropological Institute*, N.S.13(1): S19–S38.

Ingold, T (2010). *Being alive*. Routledge, London.

Vannini, P and Vannini, A (2018). These boardwalks were made for bushwalking: disentangling grounds, surfaces, and walking experiences. *Space & Culture*, 21(1): 33–45.

Vannini, P, Waskul, D, Gottschalk, S and Ellis-Newstead, T (2012). Making sense of the weather: dwelling and weathering on Canada's rain coast. *Space & Culture*, 15(4): 361–380.

3 Moved by wind and storms

Imaginings in a changing landscape

Tonya Rooney

Introduction

This chapter is dedicated to the lands and waterways of the Ngunnawal and Yuin peoples. I acknowledge these as the lands upon which this work has been undertaken. At the time of writing, these and many other lands across Australia are being ravaged by bush fires; fires made more ferocious by drought and a lack of political will to engage with the impacts of anthropogenic climate change. Breathing in the smoke from the fires, I become conscious that the thick airborne ash consists of a mix of charred earth, vegetation, wildlife and other materials, all decimated in the path of fire. As I walk and breathe, I too am earth.

Storms are often experienced as situated, local and sometimes dramatic events. Within the midst of wind or storm, an affective resonance might tingle across skin or surface, as bodies, human and nonhuman, move with and through the shifting intensities of elemental change. In the stillness that follows, earth lies exposed in the shambles and destruction of the storm's wake. With wind and storm, it seems almost obvious to say that human movement through a weather-world happens with an earth that is also always on the move. Our human bodies sense the movement of air, water and earth around us as we are moved by and with the stormy elements; sometimes literally when gusting wind behind quickens our pace or when we rush to seek cover from an unexpected downpour. We also feel our entanglement in the weather-world (Ingold, 2010) as water seeps through our clothing or wind lifts dust through the air that we inhale. At other times we may be also moved to stillness or tears such as at the smell and lightness of first drops of rain after a long period of drought or bush fire. In the aftermath of storms, we witness changes in the landscape; trees split by lightening or felled by wind, eroded channels lining a once-smooth surface or clumps of debris gathered in odd places. We notice these shifts and come to understand our own movement during such events as somehow in relation to the moving landscape. The interconnection between weather and human mobility is also noted by Vannini et al. (2012: 369) who observe that 'weather immerses us into a world of sensations that underwrite our capacity to touch, taste, see, smell, and hear place, by affording us with possibilities for movement, for action'. As I discuss below, these possibilities are not just those that arise in 'here and now' moments, but rather are generated within and across histories and futures of collective weathering.

In this chapter, I put the immediacy and localness of individual storm events aside, and instead explore how imagining stormy moments when there is no obvious storm around, can help us to grasp the ways that all human mobility (whether on foot, bike, car, boat or plane) always involves moving along with an earth that is also shifting. I situate these imaginings within the wider transdisciplinary debate on the ways that humans might live well in a time of anthropogenic climate change, influenced, for example, by the work of Donna Haraway (2016) and Anna Tsing (2015). I engage with the notion that to better appreciate human entanglement in the earth's deep geological and climatic shifts, it seems important to find ways to move past thinking about human mobility as something that happens across, on or over a seemingly static or inert earth, or as Ingold (2015) suggests, a surface that can be readily contained or covered up. Rather, the earth is complicated and the ground far more than a hardened surface over which humans venture. To simply view the earth's surface as a platform for hyper-complex and progress-driven human activity risks a human-centric positioning in relation to the earth, which is viewed as a resource for human activity with little acknowledgement of human–earth interdependencies or relations. If, from a human perspective, we can imagine *with* dramatic weather shifts and upheavals from other times and places, even in storm-free moments of relative still and calm, this may allow us to recognise the diverse temporalities and mobilities of earth and cosmos that have been on the move long before our human presence. This awareness may, in turn, help us to better navigate and imagine climate futures with nonhuman forces, lives and times.

Doreen Massey (2006), in her reflection on the mobility and the ever-evolving topologies of landscape, suggests that engaging with the temporality of place can provide an opening to knowledges and imaginations that lie beyond the immediacy of our experience or our memory. Massey also argues that thinking with the temporalities of place requires us to rethink the common positioning of time as the dominant dimension of change (also see Massey, 2005) and instead, to take a more expansive and intertwined view of landscape as in both time and space. Such a view allows us to see landscape as more than a stable surface on which changes occur over time, and rather to recognise that in 'the movement of the rocks, both space and landscape [can] be imagined as provisionally intertwined simultaneities of ongoing, unfinished stories' (Massey, 2006: 46). This more lively and entangled view of time and space can help us to imagine beyond the 'here and now' of the everyday landscapes we move through. It also helps us to recognise that the 'here' through which we walk is only as it is because of the 'outrageous specialness of the current conjunction' of the here and now (ibid: 42). To be 'here' is therefore not to invoke some grounded security upon which we can come to know and tell stories of the place but is rather to be with converging histories from other times and places. The collective weathering movement of landscape over time is in some sense the very 'ongoingness' of space and time (Neimanis and Walker, 2014; Rooney, 2019a).

Imagination is a powerful methodological tool. It can be used to reach deeper times, places and worlds that are otherwise hard to know or explore. In Doreen Massey's discussion about the tectonic movements of mountains, she notes (2006: 35) that '(W)hat is important here is not the formal knowledge ... but what one allows it to do to the imagination'. Kathryn Yusoff also draws attention to the imaginative potential of rocky formations; albeit from a different perspective to the provocation offered by Massey. Yusoff (2013: 781) suggests that 'a deep history of geologic life might well elaborate on more generative climate futures ... and offer alternative imaginaries for the inhuman forces within humanity'. The inhuman forces Yusoff refers to are the geologic forces within the very 'biopolitics of life'. Through an exploration via the geologic imagination, Yusoff invites readers to engage with the 'mineralogical dimensions of humanity' (2015: 384) and to consider in what sense humans might be constituted by non-local, geological forces through the ways these bring possibilities for 'entities to be what they are'. Yusoff is not suggesting we think of ourselves as a simple hybrid of human/ nonhuman elements that are somehow stuck together, but rather that we question the boundaries that frame 'life/nonlife' distinctions and see the inhuman as 'within the very composition of the human' (ibid). While biological forces are often acknowledged as the foundation of life and change, Yusoff (2015: 398) argues that we also need to acknowledge geology in the formation of the subject, as it is a

> participant in a mixed production that is not about addition of elements but about a configuration that is always in excess of all these elements because it is nonlocal in its material and temporal affiliations. The subject is the holding together of that gap that is stretched between the discontinuities of these ecologies; it is the gathering up of these communities into an entity that is discontinuous with itself because it contains nonlocal elements.

If, with Yusoff, we in some way think of ourselves as geo subjects – as subjects who are entangled in a multitude of shared geologies across time – then perhaps through our human and inhumanness we can also recognise ourselves as part earth and part of the earth. When storms and wind shift and shape bodies (human and nonhuman), we can sense human movement as somehow on the move with a multitude of other entities. Even in the absence of immediate local weather conditions that are wild and stormy, the insights from Massey and Yussof show that humans are still always moving with and constituted by the possibilities that stem from geological histories of deeper stormy times and places. Drawing on these ideas, and reflecting on the ways in which a weathered landscape can be understood as shaping across multiple times and places, I wonder in what sense our everyday movement can be thought of as part of the shifting earth with which we move? As we are moved by earth and weather, it seems there are a number of ways in which humans are also part of the changes that we witness, not just through the way in which our actions lead to particular climate effects or how human activity may be evidenced in future fossilised layers of earth, but because there are 'geologic corporealities' within the very composition of our humanness (Yusoff, 2015: 388–389). This possibility invites us to reflect on human relations with the earth in the here and now, as well as recognising the deeper temporal and non-local geologic histories that are in excess of and challenge the boundedness of human subjectivity.

In this chapter, I rethink everyday mobilities by recounting two imaginings of stormy events and earthly upheaval. I consider how to let the mobility of the earth move our imagination (Massey, 2016) by reflecting on storm and wind events from other times and places that emerge as human imaginings in 'here and now' moments. Through these, my hope is to show some of the ways everyday human movement inevitably has connections with the deeper histories of an ever-moving earth. The imagined wild events that I draw on come from two different points in my research, and in both instances the prevailing weather conditions were still and calm. In selecting these moments, I aim to show the ways that humans might be moved by weather outside the immediate intensity of a storm (where it is perhaps more obvious that our bodily movement will be buffeted by and with weather). I draw attention to the earthly movement from other times and places that lie within the paths that humans continue to navigate. After recounting these two imaginings, I conclude the chapter with some thoughts on how upheavals brought by wind and storm also illustrate the potential in human mobility as a form of negotiation – both physically and ontologically – with earth, bodies, weather and climate futures.

Methodologically, this chapter adopts an imaginative positioning – inspired by Massey and Yusoff – based on two 'here and now' moments from my ethnographic research on human–weather relations. First, from my empirical fieldwork, I recount the imaginings of a young child through the drawing of a storm. The child draws a stormy encounter inspired by a walk I went on with his class as part of a research project conducted with preschool children in Canberra, Australia during 2017. One key aim of the project was to better understand child–weather relations through a series of regular walks in the surrounds of the preschool. This research was undertaken with co-researcher Affrica Taylor (University of Canberra), and as described in detail elsewhere (Rooney, 2019a, 2019b; Taylor, 2019; Taylor and Pacini-Ketchabaw, 2018), placed an emphasis on noticing children's relations and connections with more-than-human worlds using audio, photo, artefacts and note-taking to record the children's worldly encounters. The child's storm drawing reveals an intermingling of memories from one of these urban lakeside walks, together with another place and time imagining of wind, sea and wild storms.

The second imagining draws on my personal reflection as an ethnographic researcher curious about the complexities of human–weather relations. I recorded this account during a two-day bush salon[1] writing event in Wee Jasper, New South Wales in 2019, during which a small group of researchers walked, wrote and collectively (re)imagined our human connections and disconnections with more-than-human worlds. Walking along a bubbling creek that seemed both timeless and yet, always on the move, and bringing to the fore my research interest in the entanglement of human–weather relations, I engaged in a process of imagining with the times and places that shaped the large rocky formations stretching above the small water course. Although seemingly grounded, secure and unmoving, these rocks hold a latent liveliness that suggest a continual shaping, rumbling movement in the aftermath of storms, wind and flood water. I describe in more detail below the positioning I adopt as I attempt to imagine-with ancient rocks and histories that shape the convergence of a brief, situated human-rock encounter. Through this, I reflect on how human mobility is always happening alongside deeper earthly movements such as earthquakes and tectonic shifts (also see Taylor and Rooney, forthcoming) and the weathering forces of wind and storm.

Imagining 1: drawing up a storm

This narrative was originally recorded as a blog post (see Rooney, 2017) and has been adapted and expanded here. A group of around ten four-year-old children and three adults had been on one of our regular walks along the lakeside not far from the children's preschool. A few days later, the children are gathered inside around several tables. As if prompted by a nautical theme of treasure maps and sunken ships, the children recall some wooden remains they had spotted submerged in a lake on their walk. 'It must be the Titanic', they had exclaimed at the time. 'See, that is one of the windows!' Now, seated at a small table with an array of coloured pens and paper, some children begin to draw from that walk. The force of an imaginary storm is conjured up on one child's page, and the outline of a boat slowly disappears from view under layers of dark, swirling wind and rain. The boy explains as he draws:

> This is the black storm
> The red is fire running about in the water.
> The storm is putting out the fire.
> See how dark it's getting.
> The water is putting the fire out.

Figure 3.1 Drawing a black storm
Source: Tonya Rooney.

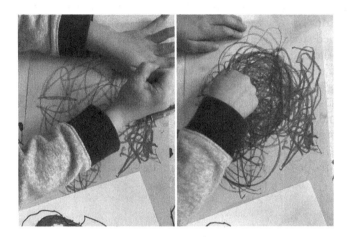

Figure 3.2 'I am drawing a storm too'
Source: Tonya Rooney.

Several other children take up this 'storm drawing', recounting stories of people going overboard, blood swirling in the waters, ships being broken to pieces and all matter reduced to nothing in the path of the dark storm.

As the imaginary storms abate, the drawings come to an end. What emerges from these drawings are not static pictures 'of' something, but stories evolving and etched in layers on the paper. These drawings are not representations of a specific place or time, but instead a compelling reminder of the liveliness of weather, place, people and the ways these might come together. Through these drawings, I am reminded of Massey's (2005) work and her hope that we might liberate our idea of space as something that is far from 'closed' or 'static' or able to be represented on the 'flat horizontality of the page'. The children's imaginings instead bring to light a notion of place as lively and ongoing, with layers of histories, temporalities and stormy happenings that find a way into and out of small everyday movements. Place in this sense can be thought of as an ever-moving 'event' that is forged in a multiplicity of relations. As Massey (2005: 140) says,

> what is special about place is precisely that throwntogetherness, the unavoidable challenge of negotiating a here-and-now ... and a negotiation which must take place within and between both human and nonhuman.

In the imagined storm drawings, temporalities are diverse and discontinuous and yet are drawn together by and through the child in ways that remind me of Yusoff's notion of human subjectivity as encompassing an excess of non-local elements. The children's drawings seem to bridge the divide between representation and event, between human and inhuman, opening an imaginary and lively space through a drawing in which they were at the same time drawn by weather. The movement in the act of drawing becomes the imagined storm, as a stormy weather event emerges from a child on the move and the child is in part the storm.

I think back to the walk by the lake that prompted these drawings. The air was calm, the sky clear and it was the very stillness of the lake's surface that allowed the children to notice planks of submerged wood under the surface.

Somewhat counter-intuitively then, without the stillness, the wild imagined storm that sank the children's Titanic, may never have surfaced or been imagined. The children's imagining navigates around any hint of a pre-given divide between earth and human movement. Rather, in walking,

Figure 3.3 'It must be the Titanic!'
Source: Tonya Rooney.

then drawing, the children imagine earth, people, fire, wind and water as moving together and moving each other. There is some irony in the children's (re)imagining of the Titanic story, given the parallels that have been drawn between this as a technological and human decision-making disaster and the progress narrative underpinning human responses to current environmental challenges (Plumwood, 2002).

On other walks undertaken as part of this research project, I also noticed the children readily traversing diverse times and scales within micro-moments of their everyday encounters, such as recognising the stillness of a caterpillar on a windy day or seeing the ongoing possibilities for life in a eucalypt tree felled by a storm (Rooney, 2019a). Consistent with these findings, in the storm drawings we see the children's capacity to imagine across and with diverse temporalities in ways that reveal storm and wind events as not something to be witnessed from outside as standalone, singular 'events' that humans only need consider in terms of their impact *on* us; rather the children are mutually becoming-with the flurry and activity of wild weather and in doing so draw attention to the temporal complexities and the liveliness of place as an ongoing 'event', a happening of histories, biologies and geologies converging to both shape and exceed the stormy moment.

Imagining 2: negotiating rocks on the move

In reiterating the ongoing movement and uncertainty in the trajectories of rocks and earth, Massey (2006: 46) observes that

> [t]he reorientation stimulated by the conceptualization of the rocks as on the move leads even more clearly to an understanding of both place and landscape as events, as happenings, as moments that will be again dispersed.

This provocation from Massey provided inspiration for a Wee Jasper Bush Salon held in September 2019, during which a small group of researchers gathered to engage in the practice of collective slow-walking and thinking-with place and multi-species others (Instone and Taylor, 2015; Taylor, 2016). Micalong Gorge, where the piece presented below was written, is located in Wee Jasper, NSW, Australia, and evokes the ancient and ongoing mobilities of rocks and landscapes. Over 400 million years ago, the rocks at Wee Jasper were laid down under the sea in a site a long way away to the south west, before arising through molten movements of volcanic eruptions and the surfacing of rocks and fossils that still lie across the landscape (Young, 2011). Imaging the deep time movement of these rocks takes us into temporal territories far before human presence on earth. Drawing on some of the walking and writing techniques that Affrica Taylor and Lesley Instone (2015) have practised, and acknowledging the influence of Kathleen Stewart's work, I experiment here with writing a narrative using the third person 'she' in an attempt to shift my positioning from a 'here and now' account, to a perspective that affords a little more distance, the significance of which I return to later in the discussion:

> With her feet, she stands and curls her toes over the edge of some rocks, trying to hold still for a moment and attune the weight and balance of her body to the undulations of the rocky terrain. The rocks feel motionless, but she can't hold still. Swaying and trying to find a point of stillness in relation to the ground below is a futile act. She gives in and imagines instead that it is the rock that is moving. Moving her. Slowly. Very slowly. She wonders how many years it would take her to get home if she moved only with the rocks. This reminds her that weather matters too; she and rock are moving today with water from the melting snow upstream. Cold and damp air swirl around. Frozen toes and the air feeling a little too close force her to break her hold with the rocks. She moves on again, walking with the creek in a time that bubbles playfully along with smaller rocks on the move. Rolling together.
>
> There is a solid precarity to the large boulders that line the gorge. It is hard to imagine them moving, but easy to see they have been moved and at any moment may move with a suddenness and force that would change everything over again. Some rocks sit perched up on the steep hill sides, while others lie crashed and cracked deep in the watery ravine. Many rocks seem to have paused, lodged together in mid-movement; but the steady flow of the mountain creek swirling around and through the rocky formations acts as a reminder that there is no stopping here. The movement is constant and relentless and always has been. Water, plants, silty and sandy earth, lichen, creek and bird life seem to grow, die, swirl and move with rocks; forcing, forging, lodging, dislodging, weathering together. Some rocks might remain in this vicinity for many hundreds of years perhaps, but they are never still and always changing.
>
> Rocks are not only tumbled and moved by the weathering together of elements, materials, beings and landscapes, they are makers and shapers of the earth – smoothing, shaping, conglomerating; folding-in and holding histories, taking and moving them onwards in time and place.
>
> She crosses a bridge over Micalong Creek. The large wooden boards have been weathered by years of vehicle crossings. Despite the splintering wood, exposed old metal bolts and a patchwork of repairs, there is something comforting about this structure. When cars cross, the clattering rhythm reminds her of a woven exchange of movement between people and other earth creatures travelling in one direction; and of rocks, water, fish, yabbies and ducks in the other, with many playful and unexpected intersections. A new sign says the bridge will be replaced soon. The bridge must be moving too. The mix of rocks and concrete that hold the bridge in place at the banks are showing signs of cracking. Nothing is really held fast for long. She imagines the volumes

Figure 3.4 Earth on the move.
Source: Tonya Rooney.

of concrete that will no doubt be brought in to build the new bridge. Crushed up rocks and sand will come from somewhere else. Rocks on the move; this time on the back of a truck.

Moving along, a windy trail leads downstream. To continue requires negotiating the slippery muddy surfaces, small patches of rocks and large wombat holes. She moves. Stepping lightly, slipping at times, reaching out with hands to steady herself on small saplings in the steeper sections. At a small clearing, she pauses, looking up at the tall Casuarinas, listening to the creek, and bending to look more closely at the micro worlds in mosses, stones and fallen trees; many now hardened, rock-like almost, by time, water, wind and weather. In this place she senses temporalities that seem deep and knowing alongside her small, not-yet-knowing and fleeting presence. She listens and moves gently with the rhythms of the undulating landscape. She wasn't invited here, and for a small moment feels some discomfort. Exposed perhaps as a human intruder, albeit well-intentioned. How to be here? How to negotiate a co-presence with the beyond human?

In this creek on the move, nothing seems too earnest. More than human companions tumble along – multiple temporalities and moving places converge and disperse – always onward. She wonders if there might be a way to recognise herself differently as she moves with the moving weathered world.

Discussion

From these two different narratives, it is possible to witness connections and other worlds as imagined and forged across time, place and stormy weather.

In the first, a child conjures up a storm that is both old and new, inspired by the stillness of a calm, urban lake. In the second, I imagine dramatic earthly movements and stormy upheavals in the deep times of the rocks on the move, reflecting on how the earth continues to shift both with and without my brief human presence; also recognising these geological movements as part of where and who I am in that moment. On the movement of rocks, Massey suggests that 'a landscape, these hills, are a (temporary) product of a meeting up of trajectories out of which mobile uncertainty a future is – has to be – negotiated' (2006: 46). From this, I wonder if the two narratives above might offer some clues into how humans might negotiate a way forward with a weathering world. Through lively and imagined events, both stories suggest a mode of negotiation that, although situated in a particular place, extends beyond 'here and now' moments of human experience. If we think of human movement as happening with a moving world rather than a kind of stepping and finding our way across a static surface, this allows us to rethink negotiation as a willingness to move and imagine with other time and places; an exchange in which we are finding a way together (Instone, 2015). In the child's drawings of the storm, humans, with all other bodies – including weather, water and ship bodies – swirl together with no normative presumptions attached to human demise; just an imagining of everything moving and weathering together. At the same time, we still witness the human part in the imagined chaos that ensues; and reflect on the responsibility that comes with being human in an entangled weather-world. In the second story, I grapple with how to negotiate a moving world underfoot; unable to hold still despite the seeming firmness of the ground and yet somehow finding footholds as I move along with the rocky and uneven terrain. The rocks also move, weathering over time as they rumble against each other in the flow of the water. In both accounts, storms and earthly movements from a distant past shape and exceed the 'here and now' and offer an imagining of being moved in a moving world.

For humans, negotiating with wind and storms is about more than seeking shelter, closing the windows, or simply waiting for a storm to abate; rather it is recognising the intermingling bodies – human and nonhuman – in a shared mobile space where new futures might be imagined. If, as Yusoff (2015) suggests, there is a sense in which humans can be understood as geological subjects, then this negotiation may also involve questioning our perception of what constitutes human substance, to recognise our geoagency and thus how our everyday mobilities might be conceived as part of the 'earth on the move'.

Both Yusoff and Massey draw our attention beyond the security of thinking of human movement as grounded in a situated and local place, by drawing attention to the deeper temporalities and geologies that lie within and exceed 'here and now' moments. This is not to deny the significance of local and situated knowledges and encounters, but rather to observe that there are relations, connections and ways we might come to know ourselves in a world that exceed the immediacy of a local experience. For Massey (2016) in particular, our movement in the local is contingent upon events beyond this. Questions of time, scale, weathering and movement are entangled in every step we take (Rooney, 2019a). The liveliness and contingency of everyday places hold significance for how we understand (human) movement in windy and stormy weather. Our entanglement with what we might perceive as a singular storm event, in fact, stretches to moments outside the storm that are still and calm, revealing a sense of the deeper histories through which we might come to understand place as an ongoing 'event' or happening. We know that humans are not only moved by weather, but that human activity in turn 'moves' the weather through the now well-documented dramatic shifts in climatic activity and wild weather events that are being witnessed across the globe. Responses of adaption and recuperation often understandably focus on present and future human–earth relations, and on the earth as something we need to fix, heal or alter. Yet arguably, unless we also recognise the deeper ways in which humans are interconnected with the geologies and meteorologies of the earth – not just in the present, but over time and place – we may limit the possibilities for living well in a weather-world with shared climate futures. Through wind and storm imaginings, human mobility can be understood as an ongoing negotiation with weather and earth; not just in terms of our physical movement but in recognition of the 'thrown-togetherness' of it all (Massey, 2005) and the geoagency of our own subjectivity (Yusoff, 2015). These concepts invite us to imagine new climate and geological futures where moving along with the world is to weather with a moving world.

Note

1 These Bush Salons are a contemporary Australian version of renowned French salons of the early 18th century that fostered women's contributions to intellectual life (Common Worlds Research Collective, 2020). Participants experiment with feminist, collaborative, creative and inclusive methods and practices that are generated in collaboration with the valley and its more-than-human inhabitants.

References

Common Worlds Research Collective 2020. *Bush Salons* [Online]. Available from: https://commonworlds.net/feminist-common-worlding-methods/bush-salons/

Haraway, D 2016. *Staying with the Trouble: Making kin in the Chthulucene*. Durham and London: Duke University Press.

Ingold, T 2010. Footprints through the weather-world: walking, breathing, knowing. *Journal of the Royal Anthropological Institute*, 16, 121–139.

Ingold, T 2015. *The Life of Lines*. Oxon: Routledge.

Instone, L 2015. Walking as respectful way-finding in an uncertain age. In *Manifesto for Living in the Anthropocene*, ed. K Gibson, D B Rose and R Fincher. New York: Punctum Books, pp. 133–138.

Instone, L & Taylor, A 2015. Thinking about inheritance through the figure of the anthropocene, from the antipodes and in the presence of others. *Environmental Humanities*, 7, 133–150.

Massey, D 2005. *For Space*. London: SAGE Publications.

Massey, D 2006. Landscape as provocation: reflections on moving mountains. *Journal of Material Culture*, 11, 33–48.

Neimanis, A & Walker, RL 2014. Weathering: climate change and the 'thick time' of transcorporeality. *Hypatia*, 29, 558–575.

Plumwood, V 2002. *Environmental Culture: The Ecological Crisis of Reason*. London and New York: Routledge.

Rooney, T 2017. Weather drawing. *Walking with Wildlife in Wild Weather Times* [Online]. Available from: https://walkingwildlifewildweather.com/2017/10/19/weather-drawing/

Rooney, T 2019a. Weathering time: walking with young children in a changing climate. *Children's Geographies,* 17, 177–189.

Rooney, T 2019b. Sticking: the lively matter of playing with sticks. In *Feminist Post-Qualitative Research for 21st Century Childhoods*, ed. D Hodgins. London and New York: Bloomsbury, pp. 43–51.Taylor, A 2019. Rabbiting: troubling the legacies of invasion. In *Feminist Post-Qualitative Research for 21st Century Childhoods*, ed. D Hodgins. London and New York: Bloomsbury, pp. 111–118.

Taylor, A and Pacini-Ketchabaw, V 2018. *The Common Worlds of Children and Animals: Relational Ethics for Entangled Lives*. London: Routledge.

Taylor, A and Rooney, T Forthcoming. Jolts from the geo-climes. In *Hacking the Anthropocene: Feminist, Queer, Anticolonial Propositions*, ed. J Hamilton, S Reid, P van Gelder and A Neimanis. Open University Press.

Tsing, AL 2015. *The Mushroom at the End of the World: On the Possibility of Life in Capitalist Ruins*. Princeton and Oxford: Princeton University Press.

Vannini, P, Waskul, D, Gottschalk, S and Ellis-Newstead, T 2012. Making sense of the weather: dwelling and weathering on Canada's rain coast. *Space and Culture,* 15, 361–380.

Young, G 2011. Wee Jasper–Lake Burrinjuck fossil fish sites: scientific background to National Heritage Nomination. *Proceedings of the Linnean Society of New South Wales*, 132, 83–107.

Yusoff, K 2013. Geologic life: prehistory, climate, futures in the anthropocene. *Environment and Planning D: Society and Space*, 31, 779–795.

Yusoff, K 2015. Geologic subjects: nonhuman origins, geomorphic aesthetics and the art of becoming inhuman. *Cultural Geographies*, 22, 383–407.

4 Walking with the rain

Sensing family mobility on-foot

Susannah Clement

Rain, rain go away,
Come again another day

Rachel (mid-40s, mother of two, married): This is like the first day it
didn't rain so we went out.
   Mike (5 years, Rachel's son): I was on my scooter.
   Susannah (researcher): That's interesting in itself, that weather really
affects you getting out of the house.
   Rachel: Oh totally, because there were a few times that we planned to
do something and we couldn't because it was too wet or windy.

(Autumn, May 2015)

This conversation took place between Rachel, her son Mike and I on a
sunny, mild autumn day at their house in 2015. Along with Mike's older
sister and father, the family were participants in a research project which
sought to explore the walking experiences and practices of families living
in Wollongong, a regional city situated on the east coast of Australia. In
speaking to parents and children about where, when and why they walked,
the weather was frequently mentioned in relation to their daily mobility.
Conversations like this one were common. Many participants, especially
mothers, mentioned that they did not particularly like walking in the rain.
This dislike seems to be corroborated by broader research into pedestrian
behaviours. Transport studies have shown that 'poor' weather conditions
result in people choosing to drive rather than walk (Mikkelsen and Christensen, 2009; Pooley et al., 2011). Public health research has also found that
'poor and extreme weather' that results from seasonal changes is a barrier
to physical activity (Tucker and Gilliland, 2007: 909; Wagner et al., 2019).
Conversely, studies also highlight that wet and rainy weather does not deter
all people from walking. As Pooley et al. (2014: 263) found, while some people were 'fair-weather walkers', others walked in rainy and windy weather,
provided they had protective clothing.

Within these walking studies, researchers and participants alike describe the absence or presence of rain using a range of vague and moralistic terms such as 'bad weather', 'good weather', 'inclement', 'nice' and 'fair'. But what do the terms actually mean? How can a weather phenomenon such as rain, a product of atmospheric shifts, the condensation of water vapour falling from the sky, be 'bad' or 'good'? What is 'nice' weather and for whom is it 'nice'? Furthermore, how might these feelings be related to the everyday mobility of walking in a particular place? For instance, an aversion to walking in the rain by Wollongong residents, a place that is temperate, warm and sunny most days of the year, might seem a bit amusing or even bemusing to those who live in places with a much more notable average rainfall.

Research underpinned by the mobilities turn has recognised walking as a socially and culturally mediated activity, highlighting the multiple styles of walking (Ingold and Vergunst, 2008; Lorimer, 2011; Middleton, 2011; Horton et al., 2014). However, I argue in this chapter that walking studies have not extended to considering the weather. While a multitude of pedestrian studies examine the experiences of walking and many more explore the 'barriers' to walking in an effort to improve the walkability of cities (Doyle et al., 2007; Pooley et al., 2014), the topic of weather is largely seen as a backdrop to pedestrian mobility. Western sciences have monopolised understandings of the weather through empirical measurements, systems of classification and forecasts. Such approaches ignore the individual or bodily sensations of how one might experience the weather beyond predetermined biological responses. Questioning the typecasting of 'fair', 'wet' and 'all-weather' walkers brings to the fore how weather, while material, is also social and made through the interplay between collective and individual discourses and embodied experiences of different types of bodies in different places, with different histories and lived experiences. As I have argued previously (Clement and Waitt, 2017, 2018; Clement, 2019), encouraging walking for families in order to create more pedestrian-friendly cities relies on knowing why and how families experience walking. Rather than categorising families as 'wet-weather' or 'fair-weather' walkers, it is important to understand that their identities are not fixed but rather emerge in relation to the material and expressive forces that are produced and felt by the sensuous body in place. Following this, I argue that the weather is important to explore in relation to family mobility and the gender, age and class politics that shape the family on-the-move.

This chapter explores the embodied experiences of mothers and children walking with rainy weather. It draws on sensory ethnographic materials gathered by and with mothers and children living in Wollongong, Australia. It advances the work of scholars who have attended to weather from a material feminist perspective (Neimanis and Walker, 2014; Neimanis, 2015; Rooney, 2018, 2019) to unpack how weather is not only 'a factor' that shapes the walking activities of participants but is deeply intertwined in the production of families' everyday pedestrian mobility and, in turn, the production of the family itself (Murray et al., 2019). Family and familial subjectivities, such as mother or child, are not given entities but always enrolled in a performative becoming. The chapter considers how the weather – and, in particular, rainy weather – comes to matter to making and remaking the family on-the-move.

To do this, I bring material feminist understandings of emotion, affect (Grosz, 1994; Ahmed and Stacey, 2001; Ahmed, 2004) and embodied, sensory and more-than-human engagements with the weather (Neimanis and Walker, 2014; Rooney, 2018) into conversation with Deleuze and Guattari's (1987) notion of assemblage. This conceives bodies as not bounded by the surface of the skin, not separate from the elements, but made in relation to embodied, emotional, affective more-than-human interactions. By taking this material feminist assemblage approach, rainy weather is understood as more than a natural occurring phenomenon to which people 'adapt'. Instead, walking with the rain is understood as a mode of becoming whereby subjectivities and collectives are constituted in relation to the *material* and *expressive limits* of *walking assemblages* made up of moving bodies, objects and nonhuman elements. In the examples explored in this chapter, it is the social unit of the family and subjectivities of mother and child that come to be made and made sense of through a walking assemblage; that is, a working arrangement of mother and child bodies, real and anticipated interactions with rain (light, heavy, cold, wet), suburban infrastructures (pavements, drains), and things brought along (umbrellas, gumboots, raincoats). It is how the material and expressive limits are reached, challenged and converted through the walking assemblage of moving bodies, objects and nonhuman elements that provides greater insight into how rainy weather comes to matter to families' everyday mobility.

From the weather 'out there' to 'weathering with'

While the sciences have monopolised understandings of the weather through the discipline of meteorology, the weather has also been a topic of investigation within the social sciences and humanities. Anthropologists Strauss and Orlove (2003) have even coined the term 'ethnometeorology' to explore how different cultures think, feel, describe and comprehend the weather and climate. Likewise, Sherratt et al.'s (2005) edited book brings together diverse meteorological, historical and anthropological accounts to position weather not only as a naturally occurring phenomenon governed by the physical sciences but as socially and culturally constructed, whereas Barnett's (2015) natural and cultural history of rain weaves together the development of meteorology, pop culture references and ancient cultural rituals. Geographers have similarly engaged with social commentaries on the weather, highlighting how sociocultural weather practices are not only social constructions but are embodied and felt (Gorman-Murray, 2010; Hitchings, 2011; de Vet, 2013, 2014; Bell et al., 2019; Simpson, 2019).

For example, the relationship between bodies, weather, climate and seasonal changes is highlighted in de Vet's (2013, 2014) doctoral study of people's everyday practices, or what she calls 'weather-ways'. Looking at the experiences of people living in tropical Darwin, Australia, de Vet (2014) argues that thermal comfort is contingent upon social practices performed in relation to the intensity and duration of weather events, often including the presence, absence or anticipation of rain.

The weather does more than just shape *our* everyday practices. Positioning weather as personal disregards how weather also shapes other non-human beings, processes, landscapes and how we might shape weather. We are always with the weather, always feeling it, always being weathered by it, always *weathering*. This follows Ingold's (2007, 2010) phenomenological approach that argues not only for a more social and embodied approach to weather but a more relational approach which sees no clear boundaries between the human body and the elements, where all actions are weather-related. In expanding on how people living on the rainy Canadian west coast skilfully weather their lives by moving alongside atmospheric patterns, Vannini et al. (2012) similarly understand bodies as entangled in and always engaged in the multisensual process of weathering.

Work by material feminist scholars has further extended this notion of weathering, arguing for a relational approach to explore how the weather matters to everyday life. Rooney (2018: 6) writes, '(W)e cannot …know the weather from the outside, because to extract ourselves from air, sunlight, wind and water would be to remove ourselves from the elements that sustain us'. From a material feminist approach, the body is both material and social: 'there is no clear boundary between the human body and the elements' (ibid). In other words, the lived body is not identical to the material entity that is bounded by the skin (McGavin, 2014). Following Ahmed and Stacy (2001), skin should be rethought not as a barrier to the outside world or a container of bodies but rather as a sensory surface that constitutes bodies, objects and places. It is through skin as a permeable surface that subjects emerge through feelings and affects (Ahmed, 2004). Building on these understandings of the body as co-constituted by its engagements with others, Neimanis and Walker use the concept of weathering to help disrupt the human-centric notion that weather 'happens to us'. As they (2014: 560) explain, 'weathering, then, is a logic, a way of being/becoming, or a mode of affecting and differentiating that brings humans into relation with more-than-human weather'. From this standpoint, our bodies and weather are co-constituted.

Why we choose to walk or not must always be considered in relation to the weather. As Rooney (2018) argues, walking is a mobility which opens ourselves up to the elements, often making us become more aware of our weathering and weathered bodies. However, opening up our bodies and walking with the weather, especially rainy weather, is not always comfortable (Rooney, 2018). As Nascimento (2019) likewise explores, our sometimes real and anticipated uncomfortable experiences with rain, wishing it to 'come again another day' as the popular children's nursey rhyme goes, are a function of the pervasive nature/culture dualism, whereby we seek separation between our bodies and the elements. Exploring the moments where rainy weather disrupts our daily plans through dampening our clothes, dripping and seeping onto our skin, is a starting point to thinking differently about our weather relations.

Embodied weathering walking assemblages

Deleuze and Guattari's (1987) notion of assemblage is useful for exploring the manifestation of intensities and flows which shape families' experiences of walking with rain. For Deleuze and Guattari (1987) assemblages are provisional working arrangements of bodies, ideas, items, feelings and movements which work to co-create particular identifiable relations in moments in time and space, such as a family walking together. It is important to note, however, that an assemblage is not merely a collection of 'things'. As Buchanan (2017) argues, assemblage is not an adjective; it is a verb. It offers a way of analysing particular sets of productive circumstances that co-produce subjects (mother, child), places (footpaths, roads), actions (a walk to school, a walk to the shops) and relations (a family). And so, whilst in this chapter I may describe the walking assemblage as a particular 'thing', it is not. It is merely a set of relations recognisable through their likeness to many other sets of relations, similar lived experiences or social norms which co-constitute bodies in familiar yet discrete moments in time and place.

Following Deleuze and Guattari (1987), assemblages are structured through *forms of content* and *forms of expression*. Forms of content, known as the machinic assemblage of bodies, are the non-discursive or the material. They are the internal limits of assemblages, that is, for example the physical limitations of a material body to walk in specific weather conditions. On the other hand, forms of expression are the immaterial, the discursive, the embodied, the affective and the social. For example, they describe the social limits that may be imposed by underlying discourse of 'appropriate' weather for walking with children, or the affective capacity of weather felt as comfortable. Assemblages form, shift, collapse and reform in relation to the interaction of these two dimensions through the ongoing process of territorialisation (Deleuze and Guattari, 1987; Aurora, 2014). The material and expressive forces are the requirements or the limits of an assemblage to take a shape or maintain itself in a recognisable form.

While the terms 'internal' and 'external' sound quite dialectic and static, the limits or conditions in which assemblages can exist are anything but (Buchanan, 2017). The forms of materials and expressions that make assemblages are independent but come to matter through their interaction during moments of encounter. For example, what it means to be and feel comfortable whilst walking with rain is not just expressive but is determined by the materiality of bodies and places. Comfort is not found within the body; rather, comfort is produced by bodies, places and objects. As Bissell explains, comfort is an 'affective resonance' (2008: 1701) that comes into being within a particular working arrangement or an assemblage of bodies in relation to place and technologies. Comfort is conceived as embodied, personal, social, immediate and anticipated, material and immaterial, discursive and non-discursive. Furthermore, achieving comfort is not the end point. Just as bodies become comfortable, they can move beyond this affective sensation and become uncomfortable. Hence, comfort is not fixed, comfort must always be maintained. Comfort is both material and expressive, it is *constituted within* assemblages and *constitutes* assemblages of bodies who walk with the rain.

The interplay between the material and expressive forces during weathering encounters can be further understood when considering the ways in which the internal or physical limits of assemblages may be extended, or rather *converted*, in order to maintain their productive working arrangement. For example, when thinking about walking in hot sunny weather, material items, such as a sunhat, may be added to the walking assemblage to maintain its working arrangement of bodies moving comfortably. The hat becomes expressive in its maintenance of comfort and upkeep of medicalised understandings of harm from hot sun. In producing these effects, the affective resonances of skin encounters with sunlight may shift from uncomfortable to comfortable. The hat converts the walking assemblage in order for it to maintain its functioning. The hat becomes an *assemblage converter.*

Focusing on what the weather does to bodies, and what bodies do to the weather, the remainder of this chapter explores how the material and expressive limits of walking assemblages play out in relation to participants anticipated and actual encounters with rainy weather.

Walking with families in Wollongong

Empirical material in this chapter comes from interviews with mothers and children participating in a study into family walking experiences and practices in Wollongong, Australia, a regional city 80 kilometres south of Sydney. Wollongong residents' experiences with the rain are in part shaped by their experiences of localised climatic and seasonal norms. Wollongong has a 'temperate' climate; due to its coastal proximity it experiences warm and often humid summers and moderate winters. The temperature almost never drops below freezing and only reaches above 40°C for a few days each year. It has an average rainfall of 1,123 mm, with precipitation occurring approximately on a third of days throughout the year (Bureau of Meteorology, 2019). To provide some context to the quotes, the details of the time of year and season that the research encounter took place are included.

The interviews were undertaken as part of a sensory ethnographic method developed for the project (Pink, 2009). This included interviews, drawing activities, walks recorded with a GoPro or an audio recorder, and go-along walking tours with families. This in situ and reflective approach to exploring families' mobility on-foot follows many other mobile methods approaches (Fincham, McGuinness and Murray, 2010), which seek to consider how often ephemeral but meaningful moments unfold on the move.

All of the quotes included in this chapter come from mothers (aged early 30s–50s) and children (aged 4–15 years). The absence of fathers in the empirical material chosen for this chapter is reflective of the heteronormative sample, the fathers' lower engagement in the project and the gendered division of care for children (Barker, 2011). In total, 16 families agreed to participate. Thirteen of these were parented by heterosexual married couples, the others were single-mother or grandmother households. Mothers were more likely to be working casually, part-time or on maternity leave in order to care for young children during the day or before and after school. Fathers who participated often had full-time work commitments and many were not at home during interview times, which were generally organised by women to suit theirs and their children's schedules.

The next three sections explore how the material and expressive limits of walking together with rainy weather emerges in relation to the constitution of familial subjectivities; specifically, mother and child subjects. Quotes have been chosen because when read through a material feminist lens, they open up a different kind of discussion about rainy weather, walking practices and family. They consider how the limits of walking together with rain are always shifting in relation to (re)making family on-the-move, real and anticipated weather conditions, and interplay between material items, urban infrastructures and bodies with the elements.

Keeping dry: the bodily limits of comfort

CHERIE: …during the week we'll go for a walk to the park, if we're not lazy or if the weather is good… But just the only limiting factor is that it has to be good weather because if it's too windy, too cold or raining, the kids won't enjoy [the walk].

(Winter, May 2015)

Here Cherie, a married mother of two in her mid-30s with a baby on the way, explains how rain disrupts the affective resonance of bodily comfort needed to make a walk to the park enjoyable for her and her children. Cherie's quote highlights how the material and expressive bodily limits of walking with her children shift relationally with the presence and absence of rain and wind. For example, bodies participating in walks planned as cordial 'family time' are anticipated to be comfortable. On this walk, the presence of rain is anticipated to feel uncomfortable and not result in happy dispositions and moments of familial togetherness usually facilitated by a walk to the park in the sunshine. The anticipated sensuous affects of rain highlight Ahmed and Stacey's (2001) argument that skin is not a barrier to the outside world, but a permeable site of encounter. Conceptualising skin as the site of weather encounters also aligns with Grosz's (1994) argument that feelings do not emerge in a vacuum but are felt on the body as a sociocultural artefact. In other words, mothering bodies and their skins are created by and produce affective intensities and flows experienced in relation to weathering encounters. Rainy weather shifts bodily dispositions and disrupts the working arrangement, the assemblage, that makes walking together to the park a possible activity.

The importance of mothers keeping children warm and dry is further highlighted by Jake (12 years). He explains that his mother Karen (early 40s, mother of four, married) drives him and his sisters to school on rainy days:

JAKE: Walking to the bus stop and stuff? Okay well we walk to the bus stop.
SUSANNAH: You walk to the bus stop?
KAREN: And home, because mum is so mean that she won't come and pick you up for that 200-meter walk.
JAKE: It's more in the morning, because we walk down the [road] to our friend's house and then we walk.
KAREN: About 500 meters if that.
JAKE: But on rainy days she drops us off and, when she's not busy.

(Summer, February 2016)

Jack and Karen's light-hearted conversation brings awareness to how motherhood is tied up in the provision of care for children (including teenage children) through transporting them to places not only safely (Dowling, 2000; Barker, 2011) but dryly. This is particularly important at the start of the school day, ensuring that children arrive at school presentable and not soaking wet. Chauffeuring children in cars, instead of letting them walk in the rain, was also discussed by Marge (early 40s, mother of two, married):

MARGE: … if the weather's not so good, then they are walking back [from swimming lessons] with wet hair and it's cold and winter and yeah, it's not good. But if they are just walking there [to swimming lessons] and it's raining well that's ok they are going to get wet anyway. … …Yeah, the weather's a factor as well.
SUSANNAH: So, do you kind of look at the forecast and then always have brollies pack[ed] and…?
MARGE: Yeah, yeah and it depends on how long we are going to be out. Like if the kids get wet and are going to be wet all day then that's not, well sort of responsible, yeah.

(Autumn, March 2015)

For Marge, becoming a 'responsible' mother requires children to be generally kept warm and dry and not exposed to the risks of rain-soaked clothes and skin. What Marge is describing here are the bodily limits or conditions under which the walking assemblage may operate.

Whilst the term 'limit' sounds quite dialectic and static, the limits or conditions in which assemblages exist are anything but. The forms of materials and expressions that make the walking assemblage possible are independent, always shifting and come to matter through their interaction during moments of encountering rainy weather. For example, Marge also highlights how the choice to walk on rainy days depends on the intensity and duration of rainfall, temperature and the type of walk. For instance, letting children walk in the cold winter rain and getting soaked through is not something a responsible mother should allow. Yet, there are also occasions when getting wet is conceived as less disruptive, as when children are on the way to swimming lessons. Marge goes on to explain further exceptions to the requirement of dry bodies in constituting the walking assemblage of mother and child bodies with rainy weather:

MARGE: It's refreshing to get out... Like sometimes when it's just been wet and they've been cooped up inside all day I go 'ok we're getting our raincoats on and our gumboots on and we'll just go walking around the street'. ... We just go in the puddles - at the end it was about jumping in every puddle we could. We got soaking wet.

(Autumn, March 2015)

In this instance, on a mild autumn day at home with nowhere in particular to be, walking in the rain with children is not considered irresponsible. Rather, the possibilities of rainy days and jumping in puddles present an opportunity to expel pent up energy and spend time together as a family. When the purpose of the walk is not to get somewhere, but to just get out of the house, getting wet is expected and part of the fun. In this moment, the process of producing the walking assemblage is different to other examples already discussed, yet what is produced is still shaped around recognisable familial relations and subjects. What happens here is the reorientating of material and expressive limits of bodies in relation to rain in order to produce the feelings of family and togetherness. What can sometimes be an uncomfortable affective encounter with wet weather, becomes a joyful transgression.

Umbrellas, gumboots and raincoats: assemblage converters

The previous examples show that comfort is not a fixed condition; it depends on the purpose of the assemblage it is operating in, the working arrangement of bodies, desires, affects and materialities. Yet even when outside in the rain, jumping in puddles, with children getting wet, the bodily limits of comfort and the interlinking parameters of responsible motherhood have not disappeared. As Bissell (2008) reminds us, comfort, or finding comfort, is an 'affective resonance' that comes into being within a particular working arrangement of bodies with technologies. For example, temporalities of the everyday, along with the addition of gumboots, raincoats and other wet-weather gear provided a conduit for many participants through which walking in the rain was felt as acceptable.

Furthermore, while walking in the rain with children was generally avoided by most participants (unless for fun), other participants more regularly walked in the rain. For Alicia (early 50s, mother of two, married) who does not drive, walking her children to school in the rain is a regular experience. While it could be argued that this family's limits of bodily comfort are somewhat different, this is not so true. Even for Alicia, keeping children's bodies and clothes warm and dry is still necessary for the waking assemblage to function:

SUSANNAH: What happens when it rains?

JACKY: I've got a raincoat in my bag.

SUSANNAH: Oh, ok and does it keep you all dry?

JACKY: Yes. Ethan's [son (3 years)] got an umbrella thingy on his stroller

SUSANNAH: Oh ok, what does your mum do, does she get wet?

ALICIA: I've got a raincoat, but my head gets wet. You've got rain hats too, and sometimes if it's really heavy you get to wear your gumboots

JACKY: And I have my raincoat and I like my umbrella because I like 'siiinn-nggging in the rain, what a glorious feeeeling...'

(Autumn, April 2015)

As Jacky (six years) describes, they still walk to school when it rains; she wears a raincoat and has an umbrella. Her brother Ethan is protected by a stroller cover, her mother Alicia has a raincoat and they all wear gumboots. The addition of material items become expressive in their ability to convert the walking assemblage and reduce anxieties about children arriving at school wet. As *assemblage converters*, raincoats, stroller covers, hats and gumboots generate surprising anticipated affective intensities, including moments of joy and playfulness. Umbrellas, rain and their associations with songs from popular culture, such as 'Singing in the Rain', open up the opportunity for Jacky to sense the city differently.

Getting soaked: everyday contingencies

Wet-weather gear serves Alicia, Jacky and Ethan well for the most part; however, on some occasions, stormy weather makes transporting children dry and safely on-foot difficult:

SUSANNAH: How did you get to school when it was so rainy?

ALICIA: I think Marcus [husband] drove Jac, on the days that he was off. I think one day we went up in the full rain gear. ... And I think we had to go home once or twice in full rain gear. Ethan and I didn't go to playgroup

SUSANNAH: It was too stormy?

ALICIA: Yeah, we decided [we're] not going to do that.

(Autumn, April 2015)

On a particularly stormy week, heavy rain was understood as a risk that disrupted the rhythms and routines of familial walking assemblage. As diligent parents following meteorological warnings, Alicia chose not to go out after seeing the forecast for heavy rain, whilst her husband Marcus (who works night shifts and is usually at work or catching up on sleep during school drop-off times) changed his routine to drive Jacky to school. This example highlights Rooney's (2018) argument that 'wild' weather conditions offer the opportunities for unpacking and cementing the contingencies of everyday life.

Therefore, rather than categorising families as 'wet-' or 'fair-weather' walkers, it is important to understand that their identities are not fixed, but rather emergent in relation to the material and expressive forces that are produced and felt by the sensuous body in space. For instance, while some families are more likely to walk in the rain than others, this does not mean that all rain is felt to be comfortable. Nor does it mean that all family members enjoy walking in the rain. Take for example, different experiences of walking in the rain for Janet (late 40s, mother of two, divorced) and her daughters Samantha (12 years) and Delila (10 years):

SAMANTHA: No, I hate walking in the rain

SUSANNAH: Oh?

JANET: No, well we'd catch the bus for that distance [to singing lessons], [but] walking into town if it's raining, we'll still go if we needed to and couldn't put it off...

SAMANTHA: I hate walking in the rain

JANET: ... it's too muddy and...

SUSANNAH: Why?

SAMANTHA: Because it might be really wet and then cold and sick and then...

(Late autumn, May 2015)

Samantha highlights the social limits of comfort and responsibility, with wet bodies understood to be more at risk of catching a cold. Janet repeats these concerns, voicing her anxiety about the material limits of their protective wet-weather gear, with gumboots only partially keeping mud and water away from the body.

Janet also highlighted how the type of water encountered on walks is important in shaping the affective resonances of bodily comfort:

JANET: This is the first time recently that it's actually rained and the round-abouts haven't been flooded. I wonder what they do to the drains of the roundabouts whether it's this one [on Throsby Drive] or the one on Mercury Street. They're [always] under water. ...the whole foot path goes under water and it gets so deep. You can't walk. Today's the first day walking along Victoria Street, that that part of the foot path is not under water. It's finally dried out. And no matter what direction we go, I find the drainage around here is very bad.... ... There's been a time here where we've walked around Victoria Street and the drain must have been blocked up by the railway bridge. It's like a wall of water just came over the top of us as a car drove past. It's ridiculous.

(Late autumn, May 2015).

Janet's reflection here shows how the regular occurrence of flooded footpaths and roads intensifies issues of pedestrian access in a city designed for cars by causing discomfort, frustration and disruption to familial walking routes and rhythms. For Janet, wading through flooded footpaths and having pooling stormwater splashed over the body is sensed as abject. This sensing is both bodily and social, triggered by a nexus of biological and discursive responses to feeling wet and knowing that stormwater may contain dirt and unknown pathogens. Following the feminist work of Douglas (1966), stormwater can be considered as 'matter out of place' as it disrupts notions of bodily boundaries (Nascimento, 2019) as well as of anticipated capacities of city infrastructure to deal with heavy rainfall and the maintenance of pedestrian mobility. Hence, encounters with stormwater, even momentarily, are felt as deeply uncomfortable by mothers and children. This feeling of discomfort disrupts the walking assemblage to a point where walking in these conditions is not considered possible.

What we can learn when walking *with* the rain

Walking is increasingly positioned as an activity that is beneficial for our physical health, our environment and our cities. Yet, despite the promotion of walking, rates are declining, children are increasingly being driven to places, and their independent mobility is steadily falling in Australia, as in many other Western countries (Carver et al., 2013; Schoeppe et al., 2016). Numerous studies explore the barriers to walking in an effort to improve the walkability of cities for adults as well as for children. Encouraging walking for families and making cities more pedestrian-friendly relies on knowing why and how families experience walking (Clement and Waitt, 2017; Clement and Waitt, 2018; Clement, 2019). This chapter has aimed to show how wet weather expands and narrows the possibilities for walking that are always tied to gendered politics and affective encounters that shape family – as a social and cultural becoming made through movement (Murray et al., 2019).

This chapter has explored how for mothers, the decision to go for a walk when it is raining is less about the physical effort involved than about the anticipation of bodily comfort and upholding notions of responsible motherhood. These performative caring responsibilities play out as unique moments in time and space, but they are also underpinned by normative notions that children's bodies are expected to be warm, not too hot, not cold, dry and free from dirt. This is even the case for those families who routinely walk in the rain – with items such as gumboots, hats and umbrellas working to, in some instances, extend the possibilities of walking with the rain as an appropriate mode of familial mobility. These items operate as assemblage converters, allowing a walk in rain to go ahead, and the walking assemblage (which seeks to produce familial relations) to be maintained. However, walking with the rain is not *always* possible. In Wollongong, stormy weather events reinforce the common notion that caring for and being mobile with children is deeply tied to driving. In these instances, material items enrolled to reduce or ease the anxieties of wet-weather encounters were not able to overcome the challenges of walking with rain. Lack of tree or building cover, overflowing drains and flooded footpaths highlight how the city often fails the needs of pedestrians to walk in comfort and safety.

In taking an assemblage approach inspired by material feminist thinking, this chapter has extended reconceptualisations of the weather as something that does not just happen 'out there' but is part of how we relate to the more-than-human world. It continues the work of scholars who have explored the body as fluid and subjectivity as emerging. Such approaches are useful for uncovering the bodily and gendered politics that shape everyday mobility. For example, assemblage thinking helps us to better understand how the unexpected moments, bodily sensations and 'contingency' plans that are learned social and cultural responses to weather events are tied to the production of individual and collective subjectivities. Looking at a moment of discomfort in an encounter with rain highlights not only the normative understandings about car culture that dominate the narrative about caring for children but also the inequitable way urban pedestrian infrastructure is designed and managed. If the goal of many mobilities scholars, city planners or new urbanists is to create more liveable and walkable cities, I argue that future mobility and transport studies could further benefit from a material feminist assemblage approach to exploring how our everyday bodily interactions with the weather come to matter to people's everyday lives.

References

Ahmed, S (2004). Collective feelings. *Theory, Culture and Society*, 21(2): 25–42.
Ahmed, S and Stacey, J (2001). *Thinking Through the Skin*. Ed. S Ahmed and J Stacey. London and New York: Routledge.
Aurora, S (2014). Territory and subjectivity: the philosophical nomadism of Deleuze and Canetti. *Minerva*, 18: 1–26.

Barker, J (2011). "Manic Mums" and "Distant Dads"? Gendered geographies of care and the journey to school. *Health and Place*, 17(2): 413–21.

Barnett, S (2015). *Rain: A Natural and Cultural History.* New York: Crown.

Bell, S, Leyshon, C and Phoenix, C (2019). Negotiating nature's weather worlds in the context of life with sight impairment. *Transactions of the Institute of British Geographers*, 44(2): 270–283.

Bissell, D (2008). Comfortable bodies: sedentary affects. *Environment and Planning A,* 40(7): 1697–1712. doi: 10.1068/a39380.

Buchanan, I (2017). Assemblage theory, or, the future of an illusion. *Deleuze Studies,* 11(3): 457–474.

Bureau of Meteorology (2019). *Climate Statistics for Australian Locations: Monthly Climate Statistics.* Bellambi. Available at: http://www.bom.gov.au/climate/averages/tables/cw_068228_All.shtml (Accessed: 10 September 2019).

Carver, A, Watson, B, Shaw, B and Hillman, M (2013). A comparison study of children's independent mobility in England and Australia. *Children's Geographies,* 11(4): 461–475.

Clement, S (2019). Families on-foot: assembling motherhood and childhood through care. In L Murray et al. (eds.), *Families in Motion: Ebbing and Flowing Through Space and Time.* Bingley: Emerald Publishing Limited, pp. 215–232.

Clement, S and Waitt, G (2017). Walking, mothering and care: a sensory ethnography of journeying on-foot with children in Wollongong, Australia. *Gender, Place and Culture*, 24(8): 1185–1203.

Clement, S and Waitt, G (2018). Pram mobilities: affordances and atmospheres that assemble childhood and motherhood on-the-move. *Children's Geographies*, 16(3): 252–265.

Deleuze, G and Guattari, F (1987). *A Thousand Plateaus: Capitalism and Schizophrenia.* Ed. B. Massumi. London and New York: Bloomsbury.

Douglas, M (1966). *Purity and Danger: An Analysis of Concepts of Pollution and Taboo.* London: Routledge.

Doyle, S, Kelly-Schwartz, A, Schlossberg, A and Stockard, J (2007). Active community environments and health: the relationship of walkable and safe communities to individual health. *Journal of American Planning Association*, 72(1): 19–31.

Fincham, B, McGuinness, M and Murray, L (2010). *Mobile Methodologies.* Basingstoke: Palgrave Macmillan.

Gorman-Murray, A (2010). An Australian feeling for snow: towards understanding cultural and emotional dimensions of climate change. *Cultural Studies Review,* 16(1): 60–81.

Grosz, E (1994). *Volatile Bodies: Toward a Corporeal Feminism.* Bloomington: Indiana University Press.

Hitchings, R (2011). Coping with the immediate experience of climate: regional variations and indoor trajectories. *Wiley Inter-disciplinary Reviews: Climate Change*, 2: 170–184.

Horton, J et al. (2014). "Walking … just walking": how children and young people's everyday pedestrian practices matter. *Social and Cultural Geography*, 15(1): 94–115.

Ingold, T (2007). Earth, sky, wind, and weather. *Journal of the Royal Anthropological Institute*, 13(1): 19–38.

Ingold, T (2010). Footprints through the weather-world: walking, breathing, knowing. *Journal of the Royal Anthropological Institute*, 16(1): 121–139.

Ingold, T and Vergunst, JL (2008). *Ways of Walking: Ethnography and Practice on Foot*. Aldershot: Ashgate.

Lorimer, H (2011). Walking: new forms and spaces for studies of walking. In T Cresswell and Merriman (eds.), *Geographies of Mobilities: Practices, Spaces, Subjects*. Farnham and Burlington: Ashgate, pp. 19–34.

McGavin, L (2014). "Why Should Our Bodies End at the Skin?" Cancer pathography, comics, and embodiment. *Embodied Politics in Visual Autobiography*, 30(1): 189–206.

Middleton, J (2011). Walking in the city: the geographies of everyday pedestrian practices. *Geography Compass*, 5(2): 90–105.

Mikkelsen, MR and Christensen, P (2009). Is children's independent mobility really independent? A study of children's mobility combining ethnography and GPS/mobile phone technologies. *Mobilities*, 4(1): 37–58.

Murray, L, McDonnell, L, Hinton-Smith, T, Ferreira, N and Walsh, K (eds) (2019). *Families in Motion: Ebbing and Flowing Through Space and Time*. Bingley, UK: Emerald Publishing Limited.

Nascimento, AD (2019). "Rain, Rain, Go Away!" Engaging rain pedagogies in practices with children: from water politics to environmental education. *Journal of Childhood Studies*, 44(3): 42–55.

Neimanis, A (2015). Weather Writing: a feminist materialist practice for (getting outside) the classroom. In P. Hinton and P. Treusch (eds.), *Teaching with Feminist Materialisms*. Utrecht: ATGENDER, The European Association for Gender Research, Education and Documentation, pp. 141–157.

Neimanis, A and Walker, RL (2014). Weathering: Climate change and the "thick time" of transcorporeality. *Hypatia*, 29(3): 558–575.

Pink, S (2009). *Doing Sensory Ethnography*. London: SAGE Publications.

Pooley, C et al. (2011). *Understanding Walking and Cycling: Summary of Key Findings and Recommendations*. Engineering and Physical Sciences Research Council. Available at: http://www.its.leeds.ac.uk/fileadmin/user_upload/UWCReport Sept2011.pdf (Accessed: 10 September 2019).

Pooley, C et al. (2014). "You feel unusual walking": The invisible presence of walking in four English cities. *Journal of Transport and Health*, 1(4): 260–266.

Rooney, T (2018). Weather worlding: learning with the elements in early childhood. *Environmental Education Research*, 24(1): 1–12.

Rooney, T (2019). Weathering time: walking with young children in a changing climate. *Children's Geographies*, 17(2): 177–189.

Schoeppe, S et al. (2016). Australian children's independent mobility levels: secondary analyses of cross-sectional data between 1991 and 2012. *Children's Geographies*, 14(4): 408–421.

Sherratt, T, Griffiths, T and Robin, L (eds.) (2005). *A Change in the Weather: Climate and Culture in Australia*. Canberra: National Museum of Australia Press.

Simpson, P (2019). Elemental mobilities: atmospheres, matter and cycling amid the weather-world. *Social and Cultural Geography*, 20(8): 1050–1069.

Strauss, S and Orlove, B (eds.) (2003). *Weather, Climate, Culture*. Oxford: Berg.

Tucker, P and Gilliland, J (2007). The effect of seasons and weather on physical activity: a systematic review. *Public Health*, 121: 909–922.

Vannini, P et al. (2012). Making Sense of the Weather: Dwelling and Weathering on Canada's Rain Coast. *Space and Culture*, 15(4), pp. 361–380.

de Vet, E (2013). Exploring weather-related experiences and practices: examining methodological approaches. *Area*, 45(2): 198–206.

de Vet, E (2014). *Weather-Ways: Experiencing and Responding to Everyday Weather.* PhD Thesis, University of Wollongong. Available at: http://ro.uow.edu.au/theses/4244/.

Wagner, AL et al. (2019). The impact of weather on summer and winter exercise behaviors. *Journal of Sport and Health Science*, 8(1): 39–45.

5 Running with the weather

The case of marathon

Jonas Larsen and Ole B. Jensen

Introduction

Mobility is not only mediated by surfaces and topologies but also by the material sensations of what Ingold (2010) terms 'weather-worlds'. People move in and through the air, sunshine, heat, rain, wind, snow, fog or icy roads. Yet the weather is largely absent as an analytic focus in mobilities studies. This reflects that the weather has been ignored in the social sciences despite its significance to much – especially outdoor – social life and dwelling (Vannini et al., 2012). It also suggests that until recently, mobilities studies have been preoccupied with 'comfortable bodies', 'sedentary affects' and 'machines' (such as cars, planes and trains) that are designed to work in all forms of weather and isolate people from direct sensuous exposure to the weather (Bissell, 2008). A car driver or train passenger is seldom affected by the weather unless it is inclement. As Ingold and Kurttila (2000: 187) write, 'people are becoming to perceive the weather less through immediate bodily experience, and more in terms of how it affects the performances of their vehicles'. The mobilities turn has not been deeply concerned with the corporeal experiences of active mobility and sport, hence the importance of the weather for mobilities has been a minor theme (but see Jones, 2017; Larsen, 2019; Newman and Falcous, 2012). However, this is changing. For instance, recent studies of bicycle mobilities discuss how cyclists are *never* oblivious to the weather as it impacts on their sensuous comfort while heavy wind, pouring rain and cool days might discourage them (Böcker et al., 2013; Nixon, 2012; Simpson, 2019). As Larsen (2020b: unpaginated) writes, 'unlike the heated and air-conditioned comfort of cars and trains that shelter people from the "excesses" of the weather-world, cyclists are more exposed to boiling hot summer days, freezing cold winter mornings, and thunderous rain throughout the year; they cannot avoid feverish sweat, painful ice-cold fingers, and being soaking wet'. Research shows that the weather is a liable taskscape that determines how much energy cyclists must put into sustaining the practice on any particular day (Nixon, 2012; Spinney, 2010).

We contribute to studies of active mobility by examining marathon running as an urban outdoor sport that takes place in the always-contingent weather-world. While the mobilities literature on running is slowly mounting (Cidell, 2014; Cook et al., 2016; Edensor et al., 2017; Latham, 2015; Larsen, 2019), few have explored endurance running as a sporting event (Edensor and Larsen, 2018). Similar to cyclists, we show that runners are uniquely immersed in the visceral realities of the weather-world from which there is no hiding. The weather is not only sensed but impacts directly on runners' somatic sensations, on their internal senses and for how long they can sustain a certain pace. We show how seasonal weather conditions training routines, particularly how race-day weather influences the atmosphere of both the race and the performance of runners. Bad weather can jeopardize months of serious training; the weather is therefore something that runners 'talk about a lot'.

The article is divided into three parts. First, we briefly theorize weather. Second, we review the existing phenomenological (Allen-Collinson et al., 2018) and science-based literature on weather and (marathon) running to establish what we know about the weathering of this practice. Third, based on our own (half-) marathon experiences, we compose evocative, vital, non-representational 'live auto-ethnographies' about how 'bad' and shifting weather conditions, and failure to dress correctly, influence the moods, expectations, body temperature and physical performance of our running bodies. It should be mentioned that Ole B. Jensen is a 'casual' runner who, for many years, has been running a couple of days a week and has participated in half-marathons. Jonas Larsen is a 'serious' runner who trains six times a week and runs at least one full-marathon a year (on the notions of 'casual' and 'serious' leisure, see Stebbins, 2001).

'Weather theory'

The weather may be thought of as a subset to the more global phenomena of the climate. A discipline such as meteorology articulates weather as a complex natural system of circulating air, water, pressure, turbulence and so on. However, this chapter examines the weather from the point of view of the 'geography closest in', that is, the sensuous human body (Longhurst, 1994). Concrete and situated weather conditions are felt in our multisensorial embodied relations to the 'outer environment'. People sense wind, sun, rain, temperature, and their complex intertwinement, as they inhabit and move through the world. As argued by Schusterman (2008: 8),

> To focus on feeling one's body is to foreground it against its environmental background, which must be somehow felt in order to constitute that experienced background. One cannot feel oneself sitting or standing without feeling that part of the environment upon which one sits or stands. Nor can one feel oneself breathing without feeling the surrounding air we inhale. Such lessons of somatic self-consciousness eventually point toward the vision of an essentially situated, relational,

and symbolic self rather that the traditional concept of an autonomous self, grounded in an individual, monadic, indestructible and unchanging soul.

Different authors discuss this complexity as a question of thinking human embodiment as 'open' to the world (Ihde, 1990; Ingold, 2007; Jensen and Vannini, 2016; McCormack, 2018; Vannini et al., 2012). Rather than subscribing to a notion of a self-contained human subject set aside from the material world or environment (what we would term an insular 'container body' perception), we lean on Merleau-Ponty's notion of 'osmosis'. Jensen (2016: 593) argues that 'osmosis' is a useful parable for understanding our open relation to the world:

> The enrolment of the human body into ... places creates complex assemblages where materialities are not just external to the human, but rather permeable as in what here will be termed a deep relationship of 'osmosis'. This is for example the case when flying a passenger aircraft with pressure cabin technology. Many will be familiar with the ear pressure and dry mouth resulting from this interface. It is an interface of osmosis where the body–world relation is much more complex that the subject–object dichotomy will allow for. Another example of the relationship of 'osmosis' is the way acceleration is felt deep in the body when we are driving a speeding car. Cycling downhill feeling topographies, wind and the corporeal and sensorial effects are also examples of how the body is in a permeable and 'open' relationship to mobilities systems and technologies.

Running is therefore a matter of responsive openness and osmosis between the body and the world. This 'exchange' is mediated by more or less specialized clothing, self-monitoring tracking devices, topographies and surfaces, traffic conditions and, finally, as now discussed, seasonality and 'weather-worlds'.

Ingold's notion of 'weather-worlds' (2010, 2015) is another way of addressing the 'small weather' that engulfs our existence. He argues that 'the experience of weather lies at the root of our moods and motivations' (2010: 131). The weather is not only perceived but, more importantly, something that we perceive in and inhabit as part of our dwelling, or inhabiting, of the open world (2007: 20). Moreover, the weather is part of the material world. As Ingold (2010: 132) writes in relation to air,

> To draw the limits of materiality around the surfaces of the landscape and artefacts would be to leave the inhabitants of the landscape and the users of artefacts in a vacuum. They would be unable to breathe... Let us, then, readmit the air as an essential material constituent of the inhabited world. This is easily done, yet is not without consequences for

the way we think about our relations with the environment. One consequence is that we can no longer imagine that all such relations take the form of interactions between persons and things, or that they necessarily arise from the conjoint action of persons and things assembled in hybrid networks. For the air is not a person or a thing, or indeed an entity of any kind, and cannot therefore comprise part of any networked assembly. It is rather, quite simply, a medium which, as Gibson pointed out, affords locomotion, respiration, and perception.

The weather is a contingent *medium* that channels or influences how always emplaced humans can move with the weather-world. Thus, weather-worlds exercise agency in affecting – with different intensity and force – the human ability to move and experience movement (Rantala et al., 2011).

The pop group 'Crowded House' sings: 'everywhere you go, you always take the weather with you'. It is a stretch to say that we 'take' the weather with us if we think of this as a willed act of human agency; however, it highlights the fact that our mobile lives are affected by micro-climates. Seeing running through the concept of osmosis means that we realize that the running body 'stretches' itself (Jensen, 2013) and extends into a hard-to-define volume and space around the body.

The geographical literature on 'volumetric' thinking (Elden, 2013; Klauser, 2010) also testifies to how running is not just about flat surfaces and upright bodies but also the immediate volumes of air and space surrounding the body. For instance, the shade of big trees or the dust from passing cars on a dirt road illustrate the need to understand the three-dimensionality and volumetric dimension of running. We also lean on McCormack's (2018) account of atmosphere; this implies an understanding that the weather is an important dimension of the situation, but also one that defies objectification. Rather, the weather is a matter of understanding different forms of 'mattering'.

Finally, it is important to stress that weather is not received passively or has direct impacts. Experiences of the weather are actively constructed and reflexively narrated through ongoing sensory weather work and interpretation (Allen-Collinson et al., 2018). As Vannini et al. (2012: 362) note, '(T)o weather is an active, reflexive, practical disposition to endure, sense, struggle, manipulate, mature, change, and grow in processes that, over time, implicate the place-making of one's dwelling'. Weather work is thus a skilful practice that people (fail to) learn through doing a specific practice. Becoming a competent marathon runner requires much sensory weather work, as we now discuss.

Marathon running as a weather-sensitive sport

Why do marathon runners talk so much about the weather? The simple answer is that running is an outdoor activity and that 'serious' marathon running (given the taxing distance) requires almost daily training for months on end in all sorts of weather. If one does not learn to cope with running in 'adverse' weather, one's body will never become marathon-ready. This signifies both the thrill and pain of marathon training. While most marathons take place in spring and autumn to avoid very high and low temperatures (and snow), such adverse weather conditions cannot be avoided (too often) when training for a marathon. Those that show up fit for a spring marathon have endured testing and troublesome cold runs, pitch dark routes and snowy, icy streets and paths that make running uncomfortable and even dangerous. Now and then, they might have sought refuge on an indoor air-conditioned treadmill to get some speedy sessions safely under their belt. Boiling hot, humid summer days where heat exhaustion and strokes are real risks trouble those that train for an autumn marathon. They learn to run early in the morning or late in the evening and adjust the intensity of their sessions to the elements-at-hand. What constitutes 'good running weather' is contingent on the specific 'taskscape' that awaits the runner on that particular day. What is considered pleasurable weather for most others might not be ideal for running.

While urban running is affected by the weather, it is not as weather-sensitive as, for example, sailing, wind surfing and climbing. While poor weather discourages many casual runners (Atkinson and Drust, 2005), the ethos within the serious running world – where windbreakers, sweat-absorbing T-shirts and specialized shoes are standard equipment – is that all sorts of weather is running weather. Sport sociologists (*and* serious runners themselves) show that endurance runners are skilled at 'weather work' (Allen-Collinson, 2018; Allen-Collinson et al., 2018), in reading how different elements affect surfaces, perceived temperatures and one's body heat (thermoception) (Hockey and Allen-Collinson, 2019):

> For runners too, we are intertwined with, and immersed in this weather-world, and often highly 'attuned' to atmospheres, including air quality ... Hence, part of the socialisation into both the cultural and physiological practices of this athletic endeavour consists of experiencing the ways in which varying temperatures and climatic conditions impact upon the running body, and understanding one's own strengths and weaknesses in relation to these impacts. Given the physically demanding nature of the sport, in hot weather a rise in body temperature occurs relatively quickly, and thermoception is a sense to which runners need to attend and become well attuned.
>
> (Hockey and Allen-Collinson, 2017: 52)

Seasoned marathon runners know how their body responds to different weather elements. They have built up the physical and emotional capacity to cope with difficult and stressful weather intensities and they know how to dress, hydrate and run according to the weather. During the race week, marathon runners obsessively check the weather forecast, fearing that a storm or extreme temperatures will jeopardize their preparation.

There is a seasonal pattern to urban marathons. Almost all are yearly events taking place at a time of the year when the weather is more or less ideal (or at least not bad) for fast competitive running. Marathon runners are obsessed with 'time' and improving their 'personal best time' (PB), so they have a predilection for fast courses. It is often possible to design such PB-friendly courses in cities with a flat topography and wide streets (Edensor and Larsen, 2018). However, the weather needs to help along or at least not disturb too much. This is the uncontrollable variable.

Sport science statistical studies show that the weather impacts on finish times and dropout rates (Spellman, 1996; Vihma, 2010; Vugts, 1997). The main factor is the air temperature; the optimal air temperature for marathon running is around 8–10°c (preferably with a light drizzle and overcast sky). The optimal temperature is relatively low because the running body quickly 'heats up' and produces much excess heat (Allen-Collinson et al., 2016; Cheuvront and Haymes, 2001). Moreover, marathon runners warm each other up as they run in proximate groups. Significantly colder, and especially warmer temperatures (particularly if the sun is out and humidity is high), are detrimental to running. Indeed, there is a correlation between increasing temperatures (above 8–12°c.) and decreases in finish times. High heat and humidity are also a health risk as dehydration and heat strokes become a common factor. Hydrating during the race and wearing very light clothing can counter some of the harmful effects of running in hot environments. However, a common (novice) mistake is to wear excessive clothing, with runners underestimating the heat production of their racing bodies. As the webpage Runner's World writes:

> Remember, no matter what the temperature says, your body is going to heat up as soon as you start moving. A solid rule of thumb: Dress like it's 15 to 20 degrees warmer than it actually is. "You should be uncomfortable standing outside as you wait for your watch to sync," says Elizabeth Corkum, a master trainer at Mile High Run Club in New York City. "If you're fine simply standing outside, the odds are good you'll overheat once you warm up into your run."
>
> (Runner's world 2019)

Clothing is a form of insulation that impairs heat transfer and sweat evaporation from the skin surface. Unless it is very cold, experienced runners run in minimal clothing and bright garments as they impose the least amount of resistance to evaporation.

This relatively low optimal temperature explains why most marathons are spring and autumn events. Yet temperatures often fluctuate from start to finish, with runners setting out in the early morning in low temperatures and finishing in hotter midday temperatures (this is particular the case for slower runners).

Moreover, while the event takes place at the same day each year, the actual weather can differ dramatically from one year to the next. Marathon runners remember what the weather was like at a particular race and how it affected their 'anticipated' pace. And they are always anxious about what weather will be thrown at them *this* particular year. As Runner's World writes:

> Fall and spring marathons are unpredictable, wondrous, and magnificent monsters. Chances of good or bad weather conditions are 50/50. Autumn races, like the New York City Marathon, are notoriously fickle: Will race day have the kind of weather you've been training in—hot? temperate?—or will early winter blindside you with frigid temps and biting wind?
>
> (Hadfield et al., 2019)

This volatility is verified in a scientific study of the weather of seven marathons from 2001 to 2010. For instance, while the mean temperature for Boston was 11.8 degrees Celsius, the temperatures fluctuated between 5.1 and 25.2 degrees Celsius from one year to the next. Chicago Marathon (with a median temperature of 7.5) was freezing cold one year (1.7) and extremely hot another year (25) (Ely et al., 2007; Knechtle et al., 2019).

Running with the weather is by no means a trivial thing. Rather, as we have illustrated briefly here, the ephemeral and unpredictable nature of the weather is one of the key challenges to marathon runners. The running body's osmotic openness to the world suggests that we need to think about the tasks and preparations for running through the analytical lens of 'weathering the running body'. We now want to move from these theoretical and conceptual framings to the actual practices of running. So, the following section is based on our auto-ethnographic accounts of half-marathon and marathon running.

Auto-ethnographic vignettes

Having discussed key aspects of how weather 'matters' in relation to marathon running, we now give fleshy examples of how the weather shapes specific experiences of marathon running. We follow the auto-ethnographic footsteps of carnal sport sociologists who study running through full-blown kinaesthetic participation, 'from' their lived bodies, internally and externally felt sensations (Allen-Collinson et al., 2016; Larsen, 2019). We also adhere to non-representational scholars who study how the haptic both include cutaneous contact with the external weather-world and the internal – or biological – world of the human body (Vannini, 2015a, 2015b). Moreover, we also follow in their footsteps by crafting impressionistic and animating vignettes that give a lively account of weather-worlds and event landscapes (Andrews, 2017; Vannini, 2015a).

"Wearing too many clothes at half-marathon in Aalborg"
(as experienced by Ole B Jensen)

"I'd been following the weather forecast for a few days up to this event. Even though – or perhaps rather because it took place on 12 May – the weather was unforeseeable. The last spring month before the meteorologists officially 'announce summer' always has the ability to surprise in Denmark. This run was no exception. The weather forecast predicted some clouds with some wind, occasional showers and temperatures around 15–18 degrees Celsius.

In the starting area, the wind was chilly, and the clouds came passing by swiftly. It was obvious that the weather (forecast) had diverse effects on the runners. I was in the segment that expected bad weather. I wore long running trousers and a long-sleeved shirt under my T-shirt with the obligatory running bib. I'm a 'cap runner'. The cap is an agile running artefact. If it gets too hot, you can quickly take it off and run with it in your hands until you find your skull-temperature ready to wear it again. This is different with the long-sleeved shirt that I came to regret putting on. In the start box it made sense to wear so many clothes since it was a bit chilly.

The starting gun was fired and we started to move along the waterfront as you do when 6,000 people have to move along the same road. So the kick-off was followed by the obligatory walking in a big, dense queue and only slowly turning into an in-between pace before reaching a level of less density and more speed. I followed the timekeeper with the 1:50 balloon thinking this would be OK given a slow training season. After a few kilometres, we reached the tunnel where the fun part repeats itself every year. People shout and wave as they run through the tunnel – a bit like kindergarten kids moving through a pedestrian tunnel. Upon surfacing on the other side, I noticed that my pace was a bit high for my estimated 1:50hr time but the time-pacing balloon keeper said that we were 'saving up' for the hilly parts. That was OK. It was around this time that I noticed that the sun was out and the pessimistic weather forecast was wrong. At this point, the combination of my high pace, the long distance and sunny weather made my long-sleeved T-shirt and trousers a poor mediator of my running. Wearing it turned out to be a bad move. I felt hot, sweaty. Wanting to keep the planned pace, I did not want to stop and take it off. It would either bring me to a halt (which by principle, I never do) or force me into a quite cumbersome act of changing while running. I endured this for a few more kilometres. But at about 10 km I capitulated, fearing that I would collapse otherwise. Still enforcing my principle of 'never stopping' I had to opt for the 'running change' which must have looked peculiar. Nevertheless, I managed to shift shirts just before the hilly part of the route. From earlier runs, I have experienced this part as easy. Not this time, though. Fatigued by the heat and my acrobatic clothing act, the hill felt really tough. I made it to the top without walking. Yet I was losing pace, dramatically. After the hill, I lost sight of the 1:50 balloon. I did not even care! All I cared about was making it to the end. It took me 2–3 kilometres to regain my breath, pace, and 'fighting mode'. I was 'digging in' (Hockey and Allen-Collinson, 2015) and my body had disposed the surplus heat whilst running in the light wind and with less clothing. A perfect weathering condition for my restitution.

I started scouting for the 1:50 balloon again. I got visual contact with it on the last part of the route, a very flat 2–3 kilometres stretch. Reignited, I almost caught up with the balloon and made it in 1:51."

"Stormy Frankfurt Marathon" (as experienced by Jonas Larsen)

"It is pitch dark outside. The road and the building are only occasionally lit up by a tram with a warning light. The hotel's thick windows isolate me from the winds and the hostile atmosphere. Leaves are thrown brutally around, and the branches of the trees do not get a second's break. The blinking lights reveal that heavy rain is falling on the road. In the last few days, we have constantly talked about the weather and checked the forecast. We have feared the sight that now met us through the window: relentless autumn rain and storms. This is a nightmare for a marathon runner as

> wind is fundamentally unfair. If it blows in a 160 arc from behind it helps, but in any other direction – 200 in total – it hinders. So 56 per cent of winds are against you and only 44 percent help you.
>
> (Anderson, 2011: 36)

The sight alone makes me tired and cold. I envy the near-perfect weather conditions earlier that year – zero wind, optimal temperatures and a gentle autumn sun – where I ran a new PB. Worse, this memory is overshadowed by my first Frankfurt Marathon some years earlier, when the wind picked up gravely two hours into the race. Despite having good legs, I suddenly felt like running into a wall at the end of the race where the many high-rise buildings made the already potent wind extra malicious. As a relatively fast runner, I was lucky to escape the wind when it picked up even more force: as we walked home in the relentless wind, I pitied those slower runners who with dead-tired bodies and limbs, struggled to make their way through the gale. And I remember how minutes later, the wind knocked over my (at the time) eight-year-old son, and not least the dramatic and extremely unpleasant plane ride home to an otherwise completely closed-down airport due to the hurricane. Like everyone else, I now hope the weather will show us some mercy over the next few hours.

My only compensation is that I have deliberately chosen a hotel a few hundred metres from the starting line to make the pre-start as stress-free as possible. I talk with my friends from my running community about the need to seek shelter from the wind by running in groups and not taking leads when faced with a stiff headwind. I stay in the warm hotel as long as possible and the cold temperature has not affected my body when the race starts.

From the 28 km sign onward, I'm on fire, increasing my speed. A new PB is on the horizon. However, as we hit the 38 km sign and the wind-whipping skyscrapers, I learn the hard way that tailwinds have propelled me forward for the last 8 kilometres. Now the tailwind is headlong, and I'm alone. At the 40 km sign I'm still ahead of my PB schedule by a couple of seconds, but the brutal wind overwhelms my energy-depleted body and I lose more than 30 seconds over the next 2 kilometres, and agony is written all over my defeated, wind-swept face."

"Freezing at Kyoto Marathon"[1] (as experienced by Jonas Larsen)

"A cold wind and snowflakes touch my skin as I walk out of the congress centre where the registration took place. While observing amazed Thai runners photographing the snow, it strikes me that I have never run a winter marathon before and that some extra layers of clothing are needed tomorrow. I'm somewhat relieved that I'm not planning to run fast tomorrow, but just want to experience the scenery and the atmosphere.

BEEP BEEP!

My alarm goes off at 5.30 am.

My sleepy eyes peep out of the window and I sigh with relief that there is no snow on the ground or wind stirring the trees. I arrive at the start area two hours before the race starts. Although it is freezing, we need to undress early as our bags must be delivered an hour before the start. In return, we are given a thin wind-resisting plastic poncho to protect our exposed 'racing bodies'. This is not enough to keep our bodies sufficiently warm for an hour and we soon become aware of the sense of thermoreception.

Despite wearing an extra old jumper, I am freezing cold. I am stuck in a long, slow-moving queue for the toilet, where my shivering body observes many other goosebump-plagued bodies that quiver and hug themselves. The seasoned marathon runner in me knows that it is better to wear light clothing and freeze a little at the beginning, because our bodies will quickly heat up once the race starts. Today, in this cold, I would like to be one of those inexperienced runners who wears excessive clothing. The freezing sensations only get worse in the windy, open football stadium where the start line begins. The sun is still subdued by the grey sky and there is no hiding from the wind. But minutes before the start, the sun finally breaks through the clouds and its warm rays reheat our bodies and lift both our mood and the atmosphere. I dispose of my poncho when the starting gun goes off.

My body temperature and the ambient temperature are still low, and combined with my moderate pace, I cannot seem to 'warm up' and my bone-frozen body overshadows other sensations during the first three or four kilometres. At one point, I grab a poncho from a waste bin. It does not take long before my body heats up and I'm running smoothly. My attention drifts away from the inclement weather and to what Leder (1990) calls my 'disappearing body', referring to the sense that a well-functioning body becomes absent. This, in turn, affords time and energy to indulge the senses: the breathtaking scenery of snow-clad mountains on the horizon, the boisterous atmosphere, the striking spectacles that enliven the streets for this special day."

Concluding remarks

Drawing on the literature on 'weather-worlds' and 'weather-skills' in the social sciences and running studies, we have discussed the role of weather in marathon running. Let us conclude here by first suggesting that the notion of 'weathering the running body' is a multisensorial and affectual process of engagement with the world. Second, this is a relation of relative openness or 'osmosis' rather than a self-contained 'container body' standing before the world. Such a Cartesian concept will not help us grasp the minute and detailed elements of running with the weather. Third, the situational interdependencies and intertwinement between human sensing and the weather is crucial to runners. Moreover, it is one of ambiguity. While runners build up weather competences and constantly try to predict or avoid inclemency or adjust their clothing and training, the weather sometimes defeats runners when they seek to control it as a 'variable'. They may misjudge it, or the weather may suddenly shift, as it is a volatile phenomenon. One may embark on a run in the sun and return home in the rain (or vice versa). The 'weathering of the running body' is a matter of distributed agency between the running body and the immediate environment. Lastly, the situation of running is one of volumes and the weather is precisely such an aerial medium that needs more than just 'flat concepts' to be comprehended.

Running with 'a sense of the weather' means choosing the right clothing for the expected weather on the day, and one's intended speed. It means carefully calibrating the preparations of sleep, rest and exercise with the intake of food, water and nutrition. Even so, if one has done all these preparatory calibrations, the run itself may offer surprises. The weather may change during the run, demanding on-the-run calibration. However, the body may also itself change temperature in more or less foreseeable ways. This is the dimension explored best via the notion of osmosis and captured by the term: 'weathering the running body'.

Further research on running and active mobility more broadly may further take our notion of 'weathering the running body' into account. More research could be done to illustrate how the weather conditions affect running practices, as well as how this relationship is negotiated through particular practices, knowledge, clothes and artefacts. The latter dimension deserves particular attention as the most recent world record testifies. Here the utilization of a new high-tech running shoe from Nike caused large public debate about how far we can accept 'assisting technologies' (the shoes in question allegedly contributed massively to the running result).

Marathon running with the weather is an interesting phenomenon in its own right, as well as a window into the body–space–movement relationship. Hence, exploring running with the weather informs our understanding of some very basic human conditions.

Note

1 This section cites previously published material (see Larsen 2020a).

Bibliography

Allen-Collinson, J (2018). 'Weather work': embodiment and weather learning in a national outdoor exercise programme. *Qualitative Research in Sport, Exercise and Health*, 10(1): 63–74.

Allen-Collinson, J, Jennings, G, Vaittinen, A and Owton, H (2018). Weather-wise? Sporting embodiment, weather work and weather learning in running and triathlon. *International Review for the Sociology of Sport*, 54(7): 777–792.

Allen-Collinson, J, Vaittinen, A, Jennings, G and Owton, H (2016). Exploring lived heat, "temperature work," and embodiment: novel auto/ethnographic insights from physical cultures. *Journal of Contemporary Ethnography*, 47(3), 283–305.

Anderson, A (2011). *Muck, Sweat and Gears*. London: Charlton Books.

Andrews, GJ (2017). From post-game to play-by-play: animating sports movement-space. *Progress in Human Geography*, 41(6): 766–794.

Atkinson, G and Drust, B (2005). Seasonal rhythms and exercise. *Clinics in Sports Medicine*, 24(2): e25–e34.

Bissell, D (2008). Comfortable bodies: sedentary affects. *Environment and Planning A*, 40: 1697–1712.

Böcker, L and Thorsson, S (2014). Integrated weather effects on cycling shares, frequencies, and durations in Rotterdam, the Netherlands. *Weather, Climate, and Society*, 6(4): 468–481.

Cheuvront, SN and Haymes, EM (2001). Thermoregulation and marathon running. *Sports Medicine*, 31(10): 743–762.

Cidell, J (2014). Running road races as transgressive event mobilities. *Social and Cultural Geography*, 15: 571–583.

Cook, S, Shaw, J and Simpson, P (2016). Jography: exploring meanings, experiences and spatialities of recreational road-running. *Mobilities*, 11(5): 744–769.

Edensor, T, Kärrholm, M and Wirdelöv, J (2017). Rhythmanalysing the urban runner: Pildammsparken, Malmö. *Applied Mobilities*, 3(2): 97–114.

Edensor, T and Larsen, J (2018). Rhythmanalysing marathon running: 'A drama of rhythms'. *Environment and Planning A: Economy and Space*, 50(3): 730–746.

Elden, S (2013). Secure the volume: vertical geopolitics and the depth of power. *Political Geography*, 34 (2013): 35–51

Ely, MR, Cheuvront, SN, Roberts, WO and Montain, SJ (2007). Impact of weather on marathon-running performance. *Medicine and Science in Sports and Exercise*, 39(3): 487–493.

Hadfield, J and The Runner's World Editors (2019). The best cold-weather running gear for every winter race. https://www.runnersworld.com/training/a20820878/what-to-wear-for-cool-weather-marathons/. Accessed 21 December 2019.

Hockey, J and Allen-Collinson, J (2015). Digging in. In W Bridel, P Markula and J Denison (eds.), *Endurance Running: A Socio-Cultural Examination*. London: Routledge, pp. 227–242.

Hockey, J and Allen-Collinson, J (2017). Running a temperature: sociological-phenomenological perspectives on distance running, thermoception and 'temperature work'. In AC Sparkes (ed.), *Seeking the Senses in Physical Cultures: Sensual Scholarship in Action*. London: Routledge, pp. 42–62.

Hockey, J and Allen-Collinson, J (2019). Distance runners as thermal objects: temperature work, somatic learning and thermal attunement. *Culture Machine*, 17: 1–18.

Ihde, D (1990). *Technology and the Lifeworld: From Garden to Earth*. Bloomington and Indianapolis: Indiana University Press.

Ingold, T (2007). Earth, sky, wind, and weather. *Journal of the Royal Anthropological Institute*, 13: 19–38.

Ingold, T (2010). Footprints through the weather-world: walking, breathing, knowing. *Journal of the Royal Anthropological Institute*, 16: S121–S139.

Ingold, T and Kurttila, T (2000). Perceiving the environment in Finnish Lapland. *Body and Society*, 6(3–4): 183–196.

Jensen, OB (2013). *Staging Mobilities*. London: Routledge.

Jensen, OB (2016). Of 'other' materialities: why (mobilities) design is central to the future of mobilities research. *Mobilities*, 11(4): 587–597.

Jensen, OB and Vannini, P (2016). Blue sky matter: towards an (in-flight) understanding of the sensuousness of mobilities design, *Transfers*, 6(2): 23–42.

Jones, P (2017). Mobile bodies. In ML Silk, DL Andrews and H Thorpe (eds.), *Routledge Handbook of Physical Cultural Studies*. London: Routledge, pp. 304–312.

Knechtle, B, Di Gangi, S, Rüst, CA, Villiger, E, Rosemann, T, Nikolaidis, PT (2019). The role of weather conditions on running performance in the Boston Marathon from 1972 to 2018. *PLoS ONE*, 14(3): e0212797.

Klauser, F (2010). Splintering spheres of security: Peter Sloterdijk and the contemporary fortress city. *Environment and Planning D: Society and Space*, 28: 326–340.

Larsen, J (2017). The making of a pro-cycling city: social practices and bicycle mobilities. *Environment and Planning A*, 49(4): 876–892.

Larsen, J (2018). Commuting, exercise and sport: an ethnography of long-distance bike commuting. *Social and Cultural Geography*, 19(1): 39–58.

Larsen, J (2019). 'Running on sandcastles': energising the rhythmanalyst through non-representational ethnography of a running event. *Mobilities* 4(5): 561–577.

Larsen, J (2020a). Marathon mobilities: a western tourist perspective on Japanese marathons. In H Endo (ed.), *Tourism Mobilities in Japan*. London: Routledge, pp. 124–137.

Larsen, J (2020b). Ups and downs with urban cycling. In O. B. Jensen, C. Lassen, V. Kaufmann, M. Freudendal-Pedersen and I. S. G Lange (eds.), *Handbook of Urban Mobilities*. London: Routledge, pp. 127–135.

Leder, D (1990). *The Absent Body*. Chicago: University of Chicago Press.

Longhurst, R (1994). The geography closest in – the body ... the politics of pregnability, *Australian Geographical Studies*, 32(2): 214–223.

McCormack, D (2018). *Atmospheric Things. On the Allure of Elemental Envelopment*. Durham: Duke University Press.

Newman, J and Falcous, M (2012). Moorings and movements: the paradox of sporting mobilities. *Sites*, 9(1): 38–58.

Nixon, DV (2012). A sense of momentum: mobility practices and dis/embodied landscapes of energy use. *Environment and Planning A*, 44: 1661–1678.

Rantala, O, Valtonen, A and Markuksela, V (2011). Materializing tourist weather: ethnography on weather-wise wilderness guiding practices. *Journal of Material Culture*, 16(3): 285–300.

Runner's world. (2019). What to wear when you're running. https://www.runners world.com/training/a20803133/what-to-wear/. Accessed 21 December 2019.

Schusterman, R (2008). *Body Consciousness. A Philosophy of Mindfulness and Some-aesthetics*, Cambridge: Cambridge University Press.

Simpson, P (2019). Elemental mobilities: atmospheres, matter and cycling amid the weather-world. *Social and Cultural Geography*, 20(8), 1050–1069.

Spellman, G (1996). Marathon running—an all-weather sport? *Weather*, 51(4): 118–125.

Stebbins, RA (2001). Serious leisure. *Society*, 38(4): 53.

Vannini, P (2015a). Non-representational ethnography: new ways of animating life-worlds. *Cultural Geographies*, 22(2): 317–327.

Vannini, P (2015b). Non-representational research methodologies. In P Vannini (ed.), *Non-Representational Methodologies: Re-envisioning Research*. London: Routledge, pp. 1–18.

Vannini, P, Waskul, D, Gottschalk, S and Ellis-Newstead, T (2012). Making sense of the weather: dwelling and weathering on Canada's rain coast. *Space and Culture*, 15(4): 361–380.

Vihma, T (2010). Effects of weather on the performance of marathon runners. *International Journal of Biometeorology*, 54(3): 297–306.

Vugts, HF (1997). The influence of the weather on marathon results. *Weather*, 52(4): 102–107.

6 Unexpected turbulence in aeromobilities

Kaya Barry

Introduction

Smooth and efficient air travel is dependent on a delicate negotiation of transport infrastructures, stratospheric conditions, weather events, seasonal temporalities, and other planetary systems through which the aircraft and passengers are navigated. Weather conditions such as turbulence – the sudden changes in an aircraft's pitch or altitude that causes bumps and jolts – can alter the route that the aircraft takes and the feeling of stability for those on-board. Turbulence is one of the main weather risks for flying at high altitude, impacting on the safety and comfort for passengers and crew. Difficult to forecast accurately, turbulence can be caused by a number of factors, such as rapid changes in wind speeds, temperatures, or the jet streams, as well as other forms of convective currents or storm activity. The most common form of turbulence, labelled 'clear air turbulence', is increasing globally and often occurs independently of storms or other weather phenomenon (Storer et al., 2018; Storer et al., 2017; Williams, 2016). As climate change alters seasonal weather patterns and globally increases storm activity, weather conditions such as unexpected turbulence will continue to alter the experiences and conditions of aeromobility.

In 2017–2018 the number of air passengers grew by over 7% with 4.1 billion passengers taking to the skies (IATA, 2018). There has been significant research and critique on the relationship between air travel and climate change, which focuses on the growth and demand for hypermobile forms of travel as well as the subsequent emissions produced by high-speed and resource-intensive travel (Bissell et al., 2016; Cohen and Gössling, 2015; Gössling and Peeters, 2007). Yet, to date, there has been scarce attention, particularly in mobilities and tourism geographies, on the role that weather plays in the cultures of air travel. A brief excerpt of an experience I had on-board a flight between two major Australian cities hints at the growing regularity of turbulence. Just before take-off, the captain's voice emanated from the speakers to introduce himself and the crew. The captain then stated in a rather stern and cautious tone:

> With these strong westerlies out of Melbourne we can expect a few bumps as we fly out. Also, with the strong westerlies that are in Brisbane

too, there will be a few bumps and jumps along the flight and on our descent into Brisbane. So... [long pause, then in a more humorous tone] Hang onto your hats when you're disembarking!

(19 August 2018, fieldwork journal)

As turbulence increases and becomes a regular aspect of the flying experience, I seek to explore how passengers might consider their flights as part of the changing weather and climate change. I also consider whether turbulent encounters will continue to be trivialised and subsumed into the overall experience of aeromobility, accepted as a necessary disturbance, as this fieldwork example shows.

This chapter explores turbulence as a weather phenomenon that may, in certain instances, be useful in sensorially attuning air travellers to an awareness of larger changing weather mobilities; the fostering of moments where an individual traveller's awareness of changing weather patterns in relationship to their mobility choices is one potential mechanism for thinking broadly about climate change and the impacts of high-speed and carbon-intensive forms of travel consumption. Such processes raise questions surrounding the potential futures of aeromobility when turbulence becomes less unexpected and part-and-parcel of the usual flying experience. And, taking this a step further, how might passengers be conditioned to this and become complacent in their contribution to climate change? To address these questions, I discuss a video artwork that I created using ethnographic and audio-visual documentation of an experience on-board a flight during a severe weather event. The video artwork (available at: https://youtu.be/vgpO0xVJ5s4) brings to the surface the complexity of movements that traverse the bodies of passengers and crew on board, the aircraft itself, the wind and turbulence through which the aircraft is flown, and a large tropical low-pressure system that affected the flight and the terrain below. Using creative modes of engaging with a variety of research data allows this exploration of the relationship between weather and aeromobility to uncover sensations and experiences that, at times, go beyond individual registers and solicit a broader attention to the entanglement of weather and mobilities.

It is important to note that throughout this chapter I use the term 'aeromobility' in preference, but at times in tandem with 'air travel', in order to specify the interconnected and transnational assemblages of the mobilities industries (air and land). This signals that aeromobilities encompass not only the instance or experience of flight but also the assemblages of passengers, crew, cargo, aircraft, global hubs, flight paths, jet streams, and so on. It is important to recognise the dispersion of these mobilities as a truly global endeavour, as people, goods, and all kinds of entities (human and nonhuman) are circulated around the earth.

The chapter unfolds in four sections. First, I survey a range of literature about the growth and demand for forms of hypermobility, in particular the touristic idealisation of air travel that is fuelled by particular sensory stimuli and aesthetics. Next, the relationship between sensation and mobility is explored to show the difficulties in tracing human encounters with movement that go beyond textual accounts. In the third section, I draw on creative documentation from on-the-ground and in-the-air during a severe tropical low-pressure system in Australia that caused significant turbulence. The artwork is used to explore how different sensations of turbulent weather might be considered to stimulate a momentary awareness of the entanglement of human, transport, and weather mobilities. This involves decentring the human from air travel and, instead, focusing on how sensations of movement might be felt and sensed across a variety of actors (human and nonhuman) in the situation. The chapter concludes by discussing how disruptive and confronting sensations encountered through aeromobilities, such as unexpected turbulence, might foreground the entanglements and impacts of increasing human mobility on large-scale weather and planetary changes. With demand growing for high-speed mobility, and weather patterns fluctuating with climate change, there is room for further scrutiny into the relationship of weather to aeromobilities and cultures of high-speed travel.

Aeromobility and changing weather

Whether we actively partake as a passenger on-board for leisure, commuting, work, or migration or we experience these hyper forms of mobility at the ground level, such as purchasing items that are delivered by 'express' from elsewhere, or simply look up to see and hear a plane flying overhead, for a lot of us, air travel has become ubiquitous to contemporary life. Considering aeromobilities as an assemblage of actors – both human and nonhuman – that enable and facilitate the movement of billions of passengers and cargo around the earth each year, necessitates the acknowledgement that broader constellations of actors can and do play a role in the experiences of air travel. Other nonhuman actors that influence patterns of mobility, such as the weather conditions that each flight travels through, might be too quickly considered as external influences on aeromobilities. Ash Amin and Nigel Thrift (2017: 42) remind us that '[T]hings move around the Anthropocene to a greatly heightened degree, whether we are talking about minerals, atmospheres, plants or animals'. This 'heightened degree' of mobility – in particular, generated by aeromobility – is driving some of the changes and emissions that are leading to climate change and, as a result, are drastically altering weather patterns and climatic conditions around the world.

Aeromobility has captivated passengers and imaginaries since the first commercial flights. The speed, ease, and allure of travelling by air has been central to the growth of air travel over the past decade, with no inclination of slowing. The vivid experiences of aeromobility continue to be used in airline marketing as a streamlined and smooth experience of flight offering clear views from above (Banister et al., 2012; Cohen and Gössling, 2015). However, these glossy and idealised aesthetics are increasingly problematic because they perpetuate a disconnect with the passenger on-board a clean, smooth aircraft and the ground and conditions below. We must remember that these aerial views, imbued with 'spectacle and beauty' (Adey et al., 2013: 2), must be considered in relation to the actions and politics that enable each flight. Scott Cohen and Stefan Gössling suggest that this 'glamorization of hypermobility has silenced the negative personal and social costs of frequent travel' (2015: 1662). Indeed, this 'dark side of hypermobility' (ibid) is perpetuated by the circulation of marketing and images that make aeromobility and resource-intensive travel so desirable (Banister et al., 2012).

Debates about climate change and the emissions produced from air travel continue to make headlines around the world. In 2018, the airline Ryanair became the tenth-biggest carbon emitter, the only non-power station company to make the top ten list (Neslen, 2019). There is also growing public momentum for movements such as the '#flyless' campaign (Caletrío, 2018) that urge people to slow down their mobility and opt for lower carbon-intensive travel. However, for particular areas in the world, notably in the Antipodean region, the relatively low cost and high speed of air travel is one of the only viable options for most long-distance tourism, and migratory or business travel. For instance, Australia has the third-busiest air route in the world between the major cities of Sydney and Melbourne (BITRE, 2018), and a large Fly-In-Fly-Out workforce that continues to drive demand for flights (see Barry and Suliman, 2019; Gorman-Murray and Bissell, 2018). In addition, it is difficult to separate out these relations because mobilities systems are not easily segregated into industry-specific or resource-demanding categories. Rather, as John Urry (2011: 68) points out, we need to consider mobilities within 'the many social practices in the contemporary world which in some way involve travel and transport'.

The desire for high-speed aeromobilities are also perpetuated by tourism cultures, and in particular, airline and tourist marketing that promises quick getaways to far-off destinations. The stark reality is that these faraway, exoticised landscapes are precisely the places that are feeling first-hand the devastating forces of increased severe weather events, such as tropical lows, cyclones and sea-level rise, the direct effects of climate change produced from carbon emissions from industries such as aviation. Yet, these relationships between aeromobilities, weather, and climate change are not always easy to discern for the individual passenger. However, Williams notes, 'it is becoming increasingly clear that the interaction is two-way and that climate change has important consequences for aviation' (2016: 1). It is precisely the increases in turbulence and other severe weather events that affect aviation that may be key to changing individual attention to this relationship.

Turbulence produces a feeling of 'high altitude aircraft bumpiness' (Storer, Williams and Gill, 2018: 2) and is difficult to forecast accurately or in advance as it can occur in areas where no storm activity is present. The most common form of turbulence that affects aviation, 'clear-air turbulence', is becoming more frequent and the cyclical nature of climate change and increased aviation (which then produces further emissions and alterations to the climate) will, geophysicists predict, cause turbulence to drastically increase over the next few decades (Williams and Joshi, 2013; Williams, 2016). Importantly, turbulence 'causes most weather-related aircraft incidents' (Sharman, Tebaldi, Wiener and Wolff, in Williams and Joshi, 2013), which can cause longer flight times as aircrafts re-route, while even small amounts of turbulence produce an unsettling experience for passengers on board.

From a passenger's perspective, weather is often considered as an external factor that impacts on flights. It is understandably hard to fathom how an individual taking one flight is integrally linked to the changing jet streams that are causing increased turbulence. But the movement of an aircraft through the upper atmospheres, in combination with fuel emissions, feed into the changes to which each flight contributes. The normality and familiarity with aeromobility (individuals are taking more flights, more often) tends to condition people to accept that these disruptive sensations and delays – such as encountering severe turbulence or other weather that impacts on flight times, comfort, and safety – are part-and-parcel of their mobility experiences. It is this familiarity with weather phenomenon in air travel, and the acceptance that bad weather will impact on the experience and comfort of flights that has habituated people to such experiences. Claus Lassen (2009: 180) observes this in an interview with a frequent flyer: 'He was then rather nervous, but today the plane can fly in air turbulence and he will not notice it, because travel has become a routine and a common practice'.

Even a little turbulence encountered intermittently on flights might seem acceptable to most passengers, especially for those who travel regularly by air. However, increasing turbulence and storm activity, in combination with predictions for more frequent incidents of severe clear-air turbulence, imply that turbulence will soon become one of the major factors affecting the experiences and cultures of air travel. These sensations of severe or violent turbulence on-board are difficult to come to grips with, especially for long-haul flights where sitting for hours on end with the seat belt sign on may be required. In this way, the links between air travel, climate change, and weather are becoming more pronounced and felt, especially, and crucially, for those partaking in regular flying.

Sensing mobilities

Scholars of mobilities and tourism geographies are well aware of the profound physiological impacts that long periods of hypermobility and transit have on human bodies (Anderson, 2013; Barry, 2016; Bissell et al., 2016), but to date there has been little attention paid to the impacts or sensations that aeromoblities produce beyond the human. This is important to consider because it necessitates the paying of attention to the entanglement of human mobility with other nonhumans and larger planetary systems. While the anthropocentric tendency is to consider aeromobility as simply a human endeavour, the fact is that all kinds of nonhumans are impacted and influenced by the systems, movements, and infrastructure that enable and sustain global forms of mobility. These nonhumans range from the migratory birds whose wetland habitats are diminishing with the expansion of mobility hubs and ports, or the materials, cargo, and viruses that are transported around the world (both purposefully and incidentally), to the shifts in global jet streams and wind currents that inform cloud formation, pressure flows, and weather patterns more broadly.

The sensations of aeromobility have been widely discussed in geography and the broader social sciences over the past few decades, alongside literature on airport atmospheres, mobilities design and aesthetics, security studies, and the 'non-places' of the airport and aircraft (Adey, 2010; Adey et al., 2013; Anderson, 2013; Augé, 1995; Barry, 2016; Cwerner et al., 2009; Jensen and Vannini, 2016; Kellerman, 2008; Salter, 2008). There have been several studies that attempt to untangle the affective experiences of air travel, such as Jon Anderson's exploration of jet lag as a state of disorientation that reveals the complexity of environmental cues and influences. In my previous work, I have suggested that long-distance mobility, particularly air travel, leads to a heightened sensation of disorientation that prompts a re-orientation in direct negotiation with the environmental influences such as the weather, climate, and geophysical location (Barry, 2016). Similarly, David Bissell, Phillip Vannini, and Ole Jensen's discussion of 'supercommuting' lifestyles reveals the manner in which the individual's habits of long-distance travel or hypermobility allows them to navigate through a range of personal, infrastructural, and environmental conditions. They suggest that this learned process of supercommuting involves a negotiation with 'risk, accidents, fatigue, stress, anxiety and discomfort with the variable atmospheres and surfaces travelled through' (2016: 807). While their empirical studies may not be specifically focused on weather or environmental influences per se, these atmospheric elements bleed into the vignettes of the mobility situations that affect how the individuals sense and respond.

One notable study by Ole Jensen and Phillip Vannini (2016: 23) speculates about what an aircraft might *feel* like as a way to investigate the 'sensory underpinnings of mobility'. Their investigation of 'how designed units themselves might "sense" movement through the "medium" of air' suggests that the ability to sense movement should be considered not as merely 'a passive register of external stimuli on the perceiving body and mind' but rather as 'an open disposition through which individuals are immersed in the currents of their movement' (ibid: 25–26). In this manner, they position the sensations experienced through the flight of the aircraft to be an assemblage of actors that co-compose the sensations in the situation.

Discussing the feel of take-off and acceleration, Jensen and Vannini merge their personal ethnographic accounts of flying in a large 737 aircraft and a much smaller and older DHC-2 aircraft. The ethnographic accounts traverse their perceptions and knowledge of the technological systems of the aircraft to sketch out the types of sensations that encompass the registers of both human (pilot and passengers) and nonhuman (aircraft and environment) mobility. Crucially, they attempt to speculate about what an aircraft might feel, going beyond a simple personification or anthropomorphic understanding of the aircraft as object and, instead, thinking about sensation and materiality through the aircraft as an actor. They suggest that they attempted to 'not quite "feel" as humans', 'but still "feel" in consequential ways' (2016: 38). With echoes of John Law's *Aircraft Stories* (2002) through which the TSR2 military aircraft is examined through an STS lens (Science and Technology Studies), Jensen and Vannini (2016: 37) suggest that in order to 'comprehend sensuous mobilities' a 'broader, more-than-human approach is required'. Rather than treating the aircraft as an object to sense through via a range of human stimuli, this more-than-human thinking opens up a range of other possibilities about how sensations and movements manifest in various actors, and how as humans we can perceive them. Although their study approaches the sensations of mobility through a flat ontology, I contend that their attempts to understand sensations of the aircraft and the mobilities that unfold as more-than-human are useful in exploring how mobilities that traverse weather events – such as extreme turbulence – might draw attention to the more-than-human sensations of weather and its impact on mobility.

Weather plays a vital role in the way that mobilities experiences are composed and sensed, and as I have emphasised, it is clear that the futures of aeromobility will be increasingly dictated by weather conditions. While studies point to the complexity of sensations that traverse sensing human bodies and the assortment of actors that converge during air travel, the 'external' cues such as the weather tend to remain on the peripheries of these empirical and ethnographic accounts. But these sensations might be at times more subtle, coming to our attention after travel, when we adjust to different time zones, stretch our legs, have to change clothes to suit a new climate or season, and so on. Turbulence felt during flight often lingers in one's body for several hours after landing, with a residual feeling of rocking and swaying, especially when laying down to rest. However, these are individualised sensations of mobility. Accordingly, I question what the sensations that move beyond human perception actually do in conceptualising the relationships of turbulence and weather to aeromobility. How might we use these sensations to allow us humans to attune to the more-than-human realms into which aeromobility launches? In this sense, the disruptive and confronting sensations of heavy turbulence might bring to the foreground the full force of weather mobilities within which air travel is bound.

'Unexpected turbulence' video artwork

There are a range of empirical and ethnographic accounts of the sensations of travel, but these often focus on the individual traveller's perspective, leaving the environment as external conditions to experiential reflections.

To attempt to counter this anthropocentric perspective, I have used creative processes to record and document a variety of sensations on-board several flights. The following discussion of a creative artwork and ethnographic account draws on extracts of data and audio-visual recordings collected during domestic flights between the cities of Brisbane and Melbourne, Australia, during 2017–2018. Recordings were taken in the airport terminals, on-board flights, and also in surrounding wetlands of the Brisbane airport. Before describing the video artwork, titled 'Unexpected Turbulence' (the video artwork is available to view online at https://youtu.be/vgpO0xVJ5s4), some context on the severe weather event is helpful.

A severe weather event occurred in Australia during February–March 2018, when an ex-tropical cyclone and another low-pressure system combined to create widespread flooding and high winds across 1,500 km of the north-eastern coast of Australia. In Brisbane, which is where the data and recordings in the video artwork took place, there was 135 mm of rain in the 24 hours leading up to this flight. Further north along the coast, in the waterways that flow down into the catchment areas that feed into Brisbane, there were rainfall totals of between 350 and 600 mm in 72 hours. With widespread flooding, strong winds, and generally extreme weather, flights in and out of Brisbane airport were delayed and extreme turbulence was expected on all flights. In the terminal in Melbourne, screens displaying multiple cancelled and delayed flights also warned of 'severe unexpected turbulence,' which flashed across repeatedly, warning passengers of bumpy flights ahead. The following notes from my fieldwork journal written after landing in Brisbane help to set the scene for the data and recordings that are used in the video artwork:

> The flight duration was just over 2 hours and 30 minutes, which is the scheduled duration for this route. For almost the entire flight the seatbelt sign was on, and no hot food or drinks were served. Come to think of it, I'm not sure that I was even offered a cold drink, because the flight was that bumpy. Upon boarding, the captain warned that extreme turbulence could be expected throughout the flight, and he cautioned 'to stay in your seat with seatbelt fastened at all times'. I sat at the rear of the plane with a window seat, 27F. All around the cabin I could hear people being sick near the end of the flight, with the rustle of paper bags and retching. I tried to block this out. I kept focus on the recording equipment (phone and audio recorder) and taking photographs. A few times, with the really big jolts, people screamed out in fear. The cabin crew kept reassuring everyone it was safe and that we were landing soon. Beside me sat a family with a young toddler, who did not have their own seat but instead sat on his mother's lap. He cried for most of the flight, but surprisingly, when it became really bumpy and we were jolted around, our bums lifting off the seats mid-flight, the kid laughed a lot. He found it fun. I glanced at his parents several times who both looked queasy too.
>
> (7 March 2018, fieldwork journal)

To attempt to de-centre the human and to further embed a consideration of aeromobility within weather (environment) conditions, I attempted to document the sensations and vibrations of air travel using some simple documentary tools: the accelerometer in my smartphone and an app interface designed for engineering testing, along with audio and photographic recordings. The use of technological interfaces that could *sense* movements occurring expands upon my own first-hand corporeal experiences of flying through extreme turbulence. Because the 'data' collected spans a range of technological, ethnographic, and audio-visual forms, attempting to express the situation and experience in textual accounts alone does not, I suggest, embellish the sensing of the relationships of weather and mobilities. To this end, I have used creative practices of manipulating the variety of data into the creation of an artwork – in this case, a short video, which both visualises and makes accessible to humans these sensations that are operating across and beyond individual sensorial registers. This is in line with the creative research methods that I have developed to study and expand on feelings and sensations of mobilities in geographical research (see Barry, 2019; Barry and Keane, 2020). This practice draws on arts-based explorations to bring together technologically enabled sensing and recording of movement that is at times beyond my own corporeal and sensory registers. Creative forms of research allow expression to move beyond simply verbal or textual accounts by utilising a range of research data to inform and shape the research process and analysis (Boyd and Edwardes, 2019; Hawkins, 2019).

Attempting to document the sensations of hypermobility from both on the ground and in the air, I sought to bring into focus the different types of vibrations and sensational dissonances that weather conditions have on experiences of aeromobility. Using my smartphone, I made recordings of the vibrations and tilt while on-board the flights, and before and after each flight in the departure terminals (Figure 6.1). The data recording of the vibrations and tilt were then adapted into graphical wave forms that appear in the video (Figure 6.2).

In addition, I recorded sound while on-board that captures the roar of the engines, the friction of the wind, the murmurs of people on board, and announcements and automated sounds played to passengers and crew through the aircraft cabin speakers. Matching up the graphical wave with the audio, so that the sounds and visuals appear in sync, and then collaging this with the photographs that I took on board (which can be heard in the audio

Figure 6.1 Vibrations recorded during flight.
Source: Kaya Barry.

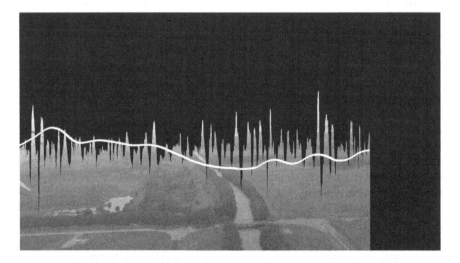

Figure 6.2 Still from the video of a flooded Brisbane landscape.
Source: Kaya Barry.

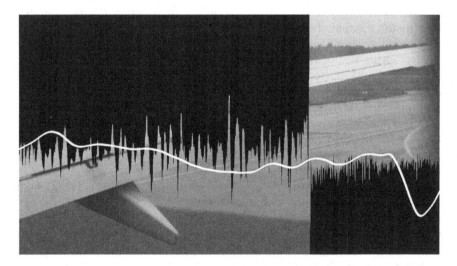

Figure 6.3 Still from video of the dramatic change in vibrations upon landing.
Source: Kaya Barry.

recording as the shutter of the DSLR camera clicks away), the video documents the approach above the water-saturated landscape and heavy turbulence and finally the landing in Brisbane airport (Figure 6.3).

Visually, the imagery and graphics on screen depict the transition from air to ground as the aircraft descends. The vibrations and tilt are significantly altered, and alongside the photographic imagery, the rough and rocky movement through turbulence leading up to the landing are re-presented. However, the creative processes that I have used in the video shift this beyond a purely representational manipulation of the various data forms, and towards a nuanced aesthetic of the multisensory forms of mobilities that the weather event produced. The sensations that occur span my body, the bodies of fellow passengers, the documentation tools, the aircraft, the navigational systems in the departure and arrival airports, the wind direction and speed, and the tropical low-pressure system. Turbulence, as one of the main weather sensations that this assemblage of hypermobility produced, caused these nauseating feelings of bouncing around, and the unsettling transition from the sunny blue skies to the descent into dark rain clouds, to finally touching down on a water-saturated landscape.

Turbulence is one type of weather that is not usually felt and accessed, or at least considered, in everyday life. While the sensations of turbulence on-board an aircraft might cause some momentary discomfort for air passengers, it is something we rarely consider from the ground. The situation that led to the recordings that I have used in this artwork involved rather extreme forms of weather wherein before take-off, the passengers of the aircraft were well aware that a potentially bumpy ride was in store. However, when 'clear air turbulence' occurs, and is predicted to increase in frequency with changing global weather patterns and climate, then this is harder for people to anticipate and rationalise. In this manner, debates on air travel need to not only engage with quantitative calculations of cause and effect (for instance, carbon emission counting) but turn towards the sensations and experiences that shape and inform global travel cultures and experiences. This is not to detract from the important work of conserving and reducing carbon emissions and waste production, nor to minimise the importance of long-distance travel in contributing to economies and, equally importantly, to maintaining social and familial connections, as more people are on the move globally.

While this artwork cannot simulate the actual experience on-board, it is important for researchers to consider how sensations influence mobility decisions, and how this might help to envisage the future effects and experiences of global mobilities. Although the video artwork that I have created is not necessarily able to re-create the weather sensations, it does draw attention to the complexity of sensations that weather produces, sensations that are often beyond human perceptions.

Conclusion: turbulent mobility futures

The Anthropocene and the drastic changes to climate and global weather patterns mean that extreme weather events will become more frequent and likely much more intense. Increasing turbulence due to severe weather and changing jet streams implies that cultures and experiences of aeromobilities will need to change to adapt and negotiate these weather events. What was previously forecast as 'unexpected' should come to be an expected and regular occurrence for experiences of air travel. Although this increased turbulence might be considered as a rather unpleasant and concerning, particularly in terms of aviation safety, aspect of the changing weather patterns, I firmly believe that more awareness and understanding of turbulence may be one way of stemming the growth and demand for high-speed travel. In this way, the result may be that the complacent attitudes towards individual consumption and desire for high-speed travel that contributes to climate change will be shaken by turbulent weather and will go beyond the generic carbon-counting discussions that are currently debated.

This chapter has explored increasing aeromobilities in relation to changing weather patterns of turbulence. I created a series of short video artworks called 'Unexpected Turbulence' that attempt to collate the various influences and sensations of weather experienced during forms of hypermobility. These sensations traverse my position in the aircraft, the human bodies around me, the aircraft itself, and the exterior environment and weather conditions that are being moved through. Sensation plays an important role in how people might think and act in response to increasingly severe weather and changing climate. Moments of disruption, disorientation, and delay are crucial to raising awareness about how human mobility is intrinsically tied to a planetary scale and is, therefore, dependent upon a range of nonhuman movements as well. As I have shown in this chapter, turbulence is one of the more unconsidered weather phenomena that impact on global mobilities – specifically commercial passenger aviation. Considering the multiplicities of weather events and the sensations it produces on human (and more-than-human) mobilities is one way in which we may be forced to think of alternatives in the turbulent futures ahead.

References

Adey, P (2010). *Aerial Life*. Chichester and Malden, MA: John Wiley and Sons Ltd.

Adey, P, Whitehead, M and Williams, A (eds.) (2013). Introduction: visual culture and verticality. In *From Above: War, Violence and Verticality*. London: Hurst and Company, 1–16.

Amin, A and Thrift, N (2017). *Seeing Like a City*. Cambridge: Polity Press.

Anderson, J (2013). Exploring the consequences of mobility: reclaiming jet lag as the state of travel disorientation. *Mobilities*, 10(1): 1–16.

Augé, M (1995). *Non-places: Introduction to an Anthropology of Supermodernity*. Trans. H John. London: Verso.

Banister, D, Schwanen, T and Anable, J (2012). Introduction to the special section on theoretical perspectives on climate change mitigation in transport. *Journal of Transport Geography*, 24: 467–470.

Barry, K (2016). Transiting with the environment: an exploration of tourist re-orientations as collaborative practice. *Journal of Consumer Culture*, 16(2): 374–392.

Barry, K (2017). The aesthetics of aircraft safety cards: spatial negotiations and affective mobilities in diagrammatic instructions. *Mobilities*, 12(3): 365–385.

Barry, K (2019). Circadian rhythms, sunsets, and non-representational practices of time-lapse photography. In CP Boyd and C Edwardes (eds.), *Non-Representational Theory and the Creative Arts*. Singapore: Palgrave, pp. 117–132.

Barry, K and Keane, J (2020). *Creative Measures of the Anthropocene: Art, Mobilities, and Participatory Geographies*. Singapore: Palgrave.

Barry, K and Suliman, S (2019). Packing for air travel: Implications of 'travelling light' for sustainable and secure aeromobilities. *Journal of Sustainable Tourism*, 28(2): 305–318.

Bissell, D, Vannini, P and Jensen, OB (2016). Intensities of mobility: kinetic energy, commotion and qualities of supercommuting. *Mobilities*, 12(6): 795–812.

Boyd, CP and Edwardes, C (eds.) (2019). *Non-Representational Theory and the Creative Arts*. Singapore: Palgrave Macmillan.

Bureau of Infrastructure, Transport and Regional Economies (BITRE). (2018). _ *Domestic Aviation Activity 2017: Statistical Report*. Canberra: BITRE. Retrieved 9 April 2018, from: https://bitre.gov.au/publications/ongoing/files/domestic_airline_activity_2017.pdf

Caletrío, J (2018). The flying less movement. *Mobile Lives Forum*, 22 March 2018. Retrieved from: http://en.forumviesmobiles.org/2018/07/19/flying-less-movement-12600

Cohen, S and Gössling, S (2015). A darker side of hypermobility. *Environment and Planning A*, 47: 1661–1679.

Cwerner, S, Kesselring, S and Urry, J (eds.) (2009). *Aeromobilities*. Oxon: Routledge.

Gorman-Murray, A and Bissell, D (2018). Mobile work, multilocal dwelling and spaces of wellbeing. *Health and Place*, 51: 232–238.

Gössling, S and Peeters, P (2007). 'It does not harm the environment!' An analysis of industry discourses on tourism, air travel and the environment. *Journal of Sustainable Tourism*, 15(4): 402–417.

Hawkins, H (2019). Geography's creative (re)turn: Toward a critical framework. *Progress in Human Geography*, 43(6): 963–984.

IATA. (2018). Traveler numbers reach new heights: IATA World Air Transport Statistics released. *International Air Transport Association*, 6 September 2018. Retrieved from: https://www.iata.org/pressroom/pr/Pages/2018-09-06-01.aspx

Jensen, OB and Vannini, P (2016). Blue sky matter: toward an (in-flight) understanding of the sensuousness of mobilities design. *Transfers*, 6(2): 23–42.

Kellerman, A (2008). International airports: passengers in an environment of 'authorities'. *Mobilities*, 3(1): 161–178.

Lassen, C (2009). A life in corridors: social perspectives on aeromobility and work in knowledge organizations. In S Cwerner, S Kesselring and J Urry (eds.), *Aeromobilities*. Oxon: Routledge, pp. 177–193.

Law, J (2002). *Aircraft Stories: Decentering the Object in Technoscience*. Durham: Duke University Press.

Neslen, A (2019). 'Ryanair is the new coal': airline enters EU's top 10 emitters list. *The Guardian*, 2 April 2019. Retrieved from: https://www.theguardian.com/business/2019/apr/01/ryanair-new-coal-airline-enters-eu-top-10-emitters-list?CMP=share_btn_tw

Salter, M (ed.) (2008). *Politics at the Airport*. Minneapolis and London: University of Minnesota Press.

Storer, L, Williams, P and Gill, P (2019). Aviation turbulence: dynamics, forecasting, and response to climate change. *Pure and Applied Geophysics,* 176(5): 2081–2095.

Storer, L, Williams, P and Joshi, M (2017). Global responses of clear-air turbulence to climate change. *Geophysical Research Letters*, 44: 9976–9984.

Urry, J (2011). *Climate Change and Society*. Cambridge: Polity.

Williams, PD (2016). Transatlantic flight times and climate change. *Environmental Research Letters*, 11(2): 1–8.

Williams, P and Joshi, M (2013). Intensification of winter transatlantic aviation turbulence in response to climate change. *Nature Climate Change*, 3: 644–648.

7 Seafarers and weather

Maria Borovnik

Introduction

Weather, ocean and seafarers are closely intermingled with each other. The wind touches the ocean waves; rain becomes seawater; waves lap against the ship and reach out to the sky; sunlight is mirrored in the ocean surface water and droplets reflect the light in rainbow colours. In contrast, heavy winds and storms might transform waves into mountainous obstacles. Seafarers' bodies and senses are entangled with the beauty and the drama caused by weather. Their working environments are almost inseparable from affective weather situations. The changing weather, ranging often between ice-cold and burning-hot, and sometimes forming wild and vicious extremes, can be challenging. Many seafarers employed on container ships, tankers or cruise ships come from tropical climates in the Philippines, Indonesia, Kiribati and Tuvalu. While extreme weather is difficult for anyone, cold weather is often the most challenging for those from the tropics.

In this chapter, I explore the ocean/weather/ship/seafarer meshwork. I delve into the dialogues within these entanglements, which are mobile, temporal, kinaesthetic, visual and audial. More particularly, I highlight how these, sometimes playful, dialogues embrace cultural memories, specifically of Kiribati and Tuvalu, when seafarers encounter weather aboard vessels. Nigel Thrift (2008: 4) suggested 'we could perhaps live in a … more playful way', a perspective that this chapter adopts without disregarding seafarers' tough working realities. I pay attention to seafarers' embodied experiences of weather, which sometimes generate moments of joy and surprise while at other times, exaggerate the rough, heavy-duty working conditions and long hours, in a comprised vessel space. While moving along the temperamental environment of the sea that picks up the everchanging sentiments of winds and waves, seafarers develop an intuition and flexibility as they constantly adjust and comply to any of these external conditions.

This chapter draws on research with merchant seafarers from the Pacific Island states of Kiribati and Tuvalu. Interviews were conducted over a period of almost twenty years from 1999, with intermittent times spent in each island state, and during a 29-day journey on a container ship. The majority of my interviews were with men: young first-time seafarers, active middle-aged ones and retired men. I also listened to the narratives of women seafarers from Kiribati, some of whom worked on container ships and others on cruise liners. During my time aboard the ship, I became increasingly interested in the affects and embodied aspects involved with merchant shipping. Yet, I noticed in my transcripts that weather was usually not a focus during interviews, although, unsurprisingly, it was frequently mentioned by many. In fact, I often quickly dissuaded someone from talking about the weather because I had wanted to understand financial aspects, the contributions of seafarers to the economic development of their home communities, and the social processes through which seafarers (re)connect with their families and communities on land. Being on a container ship changed things for me, as over a short time period I journeyed through the tropical Indian Ocean and then up across the icy, wintery Southern and Eastern China Seas. I observed the ship forging on through day and night, sometimes feeling its way through dark, foggy, narrow spaces to reach ports (Borovnik, 2017). At other times I experienced the glittering sun on iron decks and saw sweat running down deckhands' necks and overalls after painting outside in full gear in temperatures above 40 degrees Celsius. One of my journal entries describes this:

> At 3pm I have coffee with the deck crew and after this, the bosun gives me shoes that fit well, even though they are quite clunky, and a red overall that also fits (yay!). I drink some water before climbing up onto the hot, hot, hot deck!! Incredibly hot! How do the men survive working in such heat?! One of the men gives me a bucket with some rust remover (he warns me that the remover is highly toxic and at some stage offers me his glasses to protect my eyes... he is so kind!), and a brush. So, I start putting rust protection on rusty bits on deck. After that I get to take a different bucket with some cleaning liquid and the task is to rub the whole deck with a broom to remove grey spots. Then, another man hands me a hose and I get to do the best fun job of all: hosing the deck! Nice, in this heat! ... Then I walk forward (from 'aft') to join two of the men to paint the sides of the ship. ... Painting is lots of fun, although the iron deck is so hot! I cannot kneel down for long so not to burn my legs, and I'm already wearing an overall. How can we paint those lower parts when it is so hot?

These high temperatures also affected engineers and motormen working underdeck. In contrast, I observed how men took up the struggle of tying enormous ropes during wet, cold winter nights. During mealtimes, stories were shared about extreme weather events, typhoons and weather-related accidents or feelings. These observations and narratives have led me to seeing men and women working on ships as part of a mostly invisible meshwork of humans, weather and environment, which grows and is fuelled or ruptured by changes in weather, starting all over again from somewhere (a random middle) as Deleuze and Guattari (2011) explained. The ocean, or seascape, with its human and nonhuman world (Brown, 2015), is another inseparable part of this meshwork. Merchant ships continue their freight-transport commitments even after extreme weather incidents (Borovnik, 2019). Yet, this chapter focuses on seasons rather than the extreme. The next section will frame the discussion. Subsequently, I use seafarers' narratives, firstly mostly around ice and snow, to explore how 'becoming'-ocean intersects with weather and seafarers, and how this indirectly links with island home communities. Finally, I exemplify further how weather exaggerates working conditions, and yet, by sharing the physically straining labour in strong heat or cold, seafarers working in multicultural environments may find mutual understandings and will support each other across their cultural differences.

Framing ocean weather experiences

Weather is a direct, sensual, affective engagement within which people are actively involved (Martin, 2011; Vannini et al., 2011). Weather phenomena are seen by Martin (2011) as directly connected to embodied experiences; they are 'entanglements of body-with-world'. Within the intersection of weather and landscape, Ingold emphasises that the open world is 'the very *homeland* of our thoughts' (Merleau-Ponty, 1962: 214, in Ingold, 2007: S29). Out 'in the open' Ingold (2007: S29) suggests, one can be 'immersed in the weather and with the land all around us'; we not only touch a breeze but inhale it and feel it on our skin. The open world can be experienced on the sea. Steinberg and Peters (2015: 247) have emphasised that the fluidity of ocean space is different from the terrestrial landscape in a more mobile, often exciting way: 'Ocean itself ... is three dimensional and turbulent', and by encountering this fluid materiality, one enters a world of flow, characterised by liquidities and connections, a world that is constantly becoming. This world of fluid turbulence is closely connected to wind and weather, filled with temporality, possibility and uncertainty. It is a world of relationships, between different entities, chaotic, and also rhythmic (Borovnik, 2017, 2019; Steinberg and Peters, 2015: 248).

Steinberg and Peters (also Raban, 1999 and Stanley, 2016) point out that early assumptions of the sea as monotonous, flat, or even oppressive, in contrast to the furnished, variable environment of land, can be countered: the ocean, and waves, are interacting with the atmosphere – wind and weather. Consequentially, the ocean surface, the water or the making of waves, are 'shaped by the wind' and the tides, and it is this interaction and shaping that gives 'the sea not only (ever shifting) depth but also *form* – calm or angry, placid or brooding'. Anderson and Peters (2014: 11) describe how,

> The sea is obviously fluid: it is moving in terms of its location, it is unstable in terms of its form (from still calm to waves, to tides, to storm

surges and tsunamis) and changeable in terms of its chemical state (as either solid (ice), liquid or water vapor). The water world is therefore in a constant state of becoming, it is a world of immanence and transience.

This fluid ontology of the ocean is mirrored in Ingold's (2007: S32) explanation of weather as formative:

> rainfall can turn a ploughed field into a sea of mud, frost can shatter solid rocks, lightning can ignite forest fires on land parched by summer heat, and the wind can whip sand into dunes, snow into drifts, and the water of lakes and oceans into waves.

Similarly, cold air can send snowflakes across or freeze ocean surfaces into ice-shields (Vannini and Taggert, 2014). The atmosphere that creates these interactions is 'commingling' within itself – and engaging with those who are within the sea/weather/subject multiplicity. Ingold explains this atmospheric experience: 'As we touch *in* the wind, so we see *in* the sky', or we hear the world bathed in sound when it rains, our bodies are intermingled in rain (Hull, 1997). We are plunged into the mystery of the atmospheric environment as we inhabit the open world and are immersed in sunshine, rain and wind, which we respectively see, hear and touch (Ingold, 2007). This intermingling is *different* from perceiving since one is immersed in the environment rather than observing the environment as the other. Instead, commingling occurs as a 'becoming-in-the-world' (Anderson and Peters, 2014: 11). Such *becoming* (rather than *being*) in the world assumes that 'we are part of some Nature, and reciprocally, it is from ourselves that living beings and even space speak to us' (Merleau-Ponty, 2003: 206, in Connolly, 2011: 45). Focusing on this close interaction between us being an immanent part of nature, while the external nature becomes a syntonic part of us is useful in the context of seafarers (and the ship itself) experiencing wind, waves and weather. Working *with* the ocean environment is part of the seafarers' job. The ocean is intensely unpredictable and needs spontaneous adjusting and intuitive admonition. Wave movements require counterbalancing while any ship-board tasks are continued. External temperatures, shifting with seasonal zones, which ships cross within short time periods, need quick adjustments – a change of clothes and change of mindsets – and modification through technology. Throughout, the ocean nature communicates a never-seizing sense of vulnerability (Borovnik, 2019; Hutchins, 1996) (Figures 7.1 and 7.2).

This notion of becoming and immanence is central for Deleuze and Guattari (2011/1987). The connection and heterogeneity of sea/weather/subject offers the suggestion of a becoming-ocean, a becoming-weather and a becoming-subject/seafarer, all intertwined with each other. The ocean, connected within a broader network of spaces (Anderson and Peters, 2014), is surrounded by land, or becomes solid ice, and is inhabited by species who

Figure 7.1 Ships lined up waiting in front of Port of Chennai, India.
Source: Maria Borovnik.

are directed by weather. Ships and seafarers are connected within this continuingly reproducing deterritorialised ocean and weather space. Deleuze
and Guattari (2011: 238) contend that 'becoming produces nothing other
than itself', it is random. Yet, we must consider that those within the un-
bounded spaces of becoming are part of these seemingly random processes
that connect, 'grow and overspill' (p. 60), 'from a middle' (p. 21), and not from
any particular beginning or end (see also Ingold, 2011). In line with these
ideas, the *becoming ocean/weather* might turn temporarily wild, or solid, or
is interacting with land, ship and seafarers, who are surrendered by these
interactions. Langewiesche (2005: 127) describes this overpowering of ocean/
weather as an occasion on which 'the ocean seized control', as wind and a
low-pressure system take over, rattling and crashing the vessel and everyone
aboard. Ocean currents are also determined by the temperature and bar-
ometric pressure of the air surrounding water and affect both human and
nonhuman lives. Hays (2017), for example, explains that currents, which are
influenced by air temperature, may cause large detours for whales, turtles or
other migrating sea-life. Similarly, seafarers and species within the ocean are
deterritorialised in that they leave or at least traverse legal territories during
their journeys across or within ocean currents. Despite this, seafarers may

Figure 7.2 Early winter morning Port of Yantian, Republic of China.
Source: Maria.Borovnik

not become disconnected from their culture and, instead, as this chapter argues, may reconnect with their home communities throughout their weather experiences through the maintenance of their cultural values.

These conceptualisations of becoming are revisited throughout the chapter to identify the entangled phenomena of seafarers and weather: as ocean/weather/subject. By subject, I am referring to seafarers journeying across the seas while they are working on merchant vessels, such as container ships or tankers, and their connections to communities and families at home. Hence, embodied experiences with ocean/weather expand or interact over distance. In other words, seafarers, whose subjectivities originate in collective communities, experience weather in a shared and collective way.

Weather/ocean/seafarer interplay

Being intermingled with the atmosphere with its active interactions affecting waves, weather and senses, is undoubtedly experienced on ocean-going vessels – cargo ships, tankers, container ships. Here, the ship technology and constrained ship space enters this playing field of close interaction. On ships it is a combination of technological knowledge, comprehending maps showing the ocean geology, depths, latitudes and longitudes, and also a sharp intuitive sensing of the ocean, an ability to 'read the weather', and to sense any environmental oddities, that enables the safe steering of a vessel across the sea (Hutchins, 1996; Peters, 2014). Seafarers are related to the sea through the medium of the ship; rather than just seeing or hearing it, they are actually sensually involved, are a part of the ocean and weather and the interactions between both (Borovnik, 2019). All seafarers sense the changing weather and their bodies feel wind, rain or heat, either on their skin or through the ship-movements. Able-bodied seafarers who work on deck cope with changing and often unpredictable weather and temperatures most immediately. They have to adapt, sometimes only within days, between hot and freezing weather while also crossing time zones. Engineers and motormen working below deck also 'sense' the environment by noticing changes in movement or even unusual sounds, while keeping track with checking indicators and ship technology, and making sure that ship engines are smooth and in good condition.

The following exchange is with an experienced Tuvaluan bosun (boats man), who has been at work for some decades, reminiscing about a time travelling through the Arctic seas after I had asked, 'And how did you cope with the weather?'

We were there in winter. We were there... ahrh! I don't know. The first time I saw the ship got stuck by ice, you know. We were cleaning the hatch inside and it was so much loud noise like something scraping, scrrrscrrrscrrr. [We thought] 'Something is wrong. Yeah, let's climb up!' Everybody climbed up and looked: Wooah! – The ice! But we were inside the tank, inside the heads. So, we climbed up on deck to look [for] what is wrong. We'd look over the side. We saw ice. And then it got stuck. When it got stuck, we put the ladder down. I put the rope [down] and we climbed down on the ice. [Laughs].
Wouldn't that have been dangerous at all?
I don't know. But for us, nobody... I think it was very thick, the ice. The ship got stuck. I think only [for] one day. Mostly they have an icebreaker. An icebreaker is like a barge with a very strong bow. And you call to shore 'I am stuck here'; and then he said: 'Okay! What's your position there? Okay! That time the ice breaker will come to you. The icebreaker will come, breaking the ice until... Then it comes past your ship stuck here, breaking it up and then you go behind. ... [Once your ship is un-stuck] you go to the other ship that's stuck'. The ice breaker can take four ships.
A row of ships will follow the icebreaker. That's so interesting. This must have been quite adventurous for you guys?
Yeah! You see the captain was looking: no crew, eh?! Everybody is on the ice! 'Hey, what you doing?!' 'Captain, no more work, now we're going to walk on the ice'. So, no job that day. We only walk on the ice. Oh, I wish our people at home could see us walking the ice!
Did you take photos? To show your parents?

After that we got cameras but a lot of time later. Then they [parents] saw all the snow on the ship, the ice on the ship, eh?! Ah, they said, 'wow!' But this first experience was very nice. I said: 'now we are walking on the sea'. 'What?! You are walking on the sea?' 'Yes, but it's blocked, in ice, you know, that's why we're walking on the sea.' We saw people fishing, you know? Make a hole in the ice, umbrella, and a chair, few beers.

I heard that's the way they fish in Scandinavia. Did you have an Ordinary Seaman on this ship?

We had two OS [ordinary seamen]. I call them. They are good. Only cold, eh?! I told them 'if you are cold, we put newspaper in our boots', something to ease the cold.

Within the dark and confined space, within the hatches of a tanker, this experience was first distinguished as a *sound*: a loud, scratchy noise. Instantly the men *knew* that there was something wrong and decided to climb up and get out into the open. They (and the ship) were stuck; but more importantly there is astonishment – 'the ice!' – a new, visual experience for many. And their reaction was to climb down and get to know the ice through direct interaction. Even though the ship was stuck – affected by a 'vicissitude of experience' (Connolly, 2011) caused by the changing consistency of water through cold weather – this experienced Tuvaluan bosun, together with others working on the tanker, was excited not only about the ice (the unknown at the time) but also about the actions that had to be taken: to radio ashore and arrange for an icebreaker – a new practice for him. Here then, it is the interruption and not the movement itself that bears 'life as potential' (Thrift, 2008: 5). Men are able to walk on the ice, the transformed water, and it is even possible to go beyond this by digging a hole and trying out fishing. The description of the chair and a few beers while ice-fishing offers a joyful interruption to an otherwise intense job. Within these experiences, almost instantly, a sense of community is summoned up. First, the icebreaker supports up to four ships and vessels that have to work together to get out of trouble. Second, as it is best practice in Tuvaluan culture where the elders advise and look after their whole community, this bosun called, checked and advised the youngest on board, the most inexperienced ordinary seamen (OS). Third, this experience needed to be shared with family at home. This interaction with the unfamiliar ice and snow evoked cultural, communal values. Those at home could not possibly be left out; those in need of guidance and protection needed looking after.

Further exploring snow as a new experience of weather, the next extract also shows interaction within the kinaesthetic and visual of weather and exemplifies the strong cultural aspect of sharing. Another older, still active, Tuvaluan seafarer explains how he tried to describe snow to his parents but also – at the same moment of talking about snowflakes, he remembered child poverty in India:

I wrote [to] them about the snowflakes.
Snowflakes! Did you describe the snowflakes?

Yes. Coming from the sky. And I told them about a lot of children without homes I see.

Have you seen that? Whereabouts?

In India, I saw too many. My father said, why don't you bring them home here? We have lots of food here! I said, 'Dad, I'm not talking about four or five. There are millions!

...

What did [your parents] say about the snowflakes?

They said, oh why don't you put them in the letter and give us? [laughs]. I say 'flakes', also there is no word for flakes in Tuvaluan, eh?!

How do you describe them?

I say 'just flying, very light, and you feel nothing, but it melts'. They don't comprehend, but luckily they were in Ocean Island [now Banaba Island, which had more facilities than Funafuti Island at the time]. They see sometimes movies, you know?

New experiences of weather, such as snowflakes, install the wish in a Tuvaluan seafarer to share this with family back home. This man's family is interested in a tangible experience: could they touch these snowflakes, could he send them home? A memory about a similar incident is triggered, that of wanting to have hungry children sent home so that the family could care and feed them. This suggestion exemplifies the enormous generosity of sharing in the islands, which goes well beyond borders, even embracing another culture, and also incomprehensibility. In the Deleuzian sense, a snowflake experience is 'coded' and contains a 'milieu of exteriority' as it cannot be shared across climates (Deleuze and Guattari, 2011: 54). This seafarer realises the impossible – the millions of hungry children, and the ephemeral – the lightness and melting of snowflakes. Yet, foregrounded in the above descriptions is the collective, communal identity in which Tuvaluans and I-Kiribati individuals grow up. Through evolving in a collective understanding of oneself, there is a need to share and to care for others, and not keeping exciting experiences to oneself.

A young I-Kiribati woman, who worked on a cruise liner travelling to Alaska, describes this same icy region as her favourite part of the journey. She loved the scenery, including the views of snow on the mountains, and observing sea-life. She enjoyed these different ocean and weather experiences, although she found it difficult to deal with the cold. This following quote touches on the ocean/weather and becoming-ocean idea as the woman interchanges with some of the ocean life:

But when I went to Alaska I loved it but I couldn't take that. It was too cold.

It was too cold, but why did you love it?

But it was, what I loved is the mountains with snow on top of them. They looked so cute.

Really?

It was nice. Really loved it.

So, but you didn't go on shore there.

I did go on shore. But I couldn't stay for long because I couldn't take the... it was too cold, yeah.

And also probably really dark in winter?

Oh when we went there it was in the... in summertime.

Oh, you were there in summertime. It was always daylight.

Yeah, yeah.

That must have been strange.

Oh it was nice. That was the first time that I saw a whale in my life.

A whale?

A whale.

A whale, in Alaska.

Yeah. I remember my supervisor called me on the phone, just early in the morning, before we knocked off. He told me hey where're you. It was, I was assigned in the spa. I'm in the spa. I wanted to go to the starboard side to see something because I always tell them, all my colleagues in the spa, I said please if you see one just let me know 'cause I really want to see it with my own eyes. So, when he told me to go, I just ran. 'Where?' 'Right there! Just look up and you will see'. When I saw it jumping, I was still talking with him on the phone, and when I saw the whale jumping I shout. I just screamed: 'Whaaah!', and he was like, 'Why did you shout in my ears?!' I said 'Sorry! Oh my goodness!' You know what I [said was]: 'I never seen a whale in my life, and this is the first time I saw it'. Right after then when I knocked off I went straight... bought my phone card and I called back home and I told my Dad I saw a whale.

Whale watching is seasonal. Influenced by 'the magnetic field relative to the sun' (Greenfieldboyce, 2011), whales migrate over vast distances. Humpback whales, for example, swim up to seven thousand miles between warmer breeding grounds in Mexico or Hawaii, but not as far south as the equatorial Kiribati, towards the Arctic North. Whales are in Alaskan seas only during summer, usually from about May to October (Alaska Shore Excursions, 2019). However, as the above interview showed, for this woman who is usually acclimatised to tropical Pacific weather, Alaska is still rather 'cold' even during the summer season. There is an invisible 'line of becoming' within this interaction between the seafaring woman, the sensations of feeling cold, and the whale jumping, a transverse connection that runs 'in between' them (Ingold, 2011: 83). Ingold quotes Deleuze and Guattari (2004 [2011]: 323) in that 'a line of becoming ... passes between points it comes up through the middle' and that 'a line of becoming has neither beginning nor end', it always occurs 'in the middle: one can only get it by the middle'. In this example, the whale, in its migration north, may have journeyed along a straight line (as Greenfieldboyce, 2011 suggests), until particular changes in weather that influenced currents might have caused spiral-type detours (Hays, 2017). Meanwhile, the cruise ship has followed its own schedule, circulating through the Arctic. Incidentally, unplanned, therefore through a 'middle', both lines have entwined, and both the whale and woman jumped, for different reasons and in different spheres; yet, their excitement was knowingly or unknowingly entangled. This momentarily enmeshed line of interaction somewhere in the Arctic was then directed further, towards the woman's equatorial Pacific family in Kiribati. This woman drew on her collaborative communal upbringing in asking her multicultural cruise-ship colleagues community to let her know when a whale turned up. She even stayed on the phone with her supervisor during the experience and almost instantly called her father at home in Kiribati, not shying away from the expense of an on-board phone call, to tell him about this exciting, new interchange.

Weather, stress and resilience

Weather is not usually exciting for seafarers. Instead, it intensifies bodily felt stress because of long working hours, the extreme physical work and the constraints of the ship environment which has to be endured by Tuvaluans and I-Kiribati for a year, and in earlier times for as much as two years. Life for hard-working seafarers above or below deck, different from passengers cruising along, is not really 'slowed and decelerated' (Symes, 2012; Vannini, 2013); rather, it is accelerated and challenged when occasional extreme weather events occur (Borovnik, 2019; see also Langewiesche, 2005; and George, 2013). Having to work close to 40 degrees Celsius on iron surfaces in full safety gear as mentioned above, may be intense. However, working in freezing, minus Celsius degrees temperatures can be dangerous. Seafarers are prone to potentially slippery surfaces or a breaking rope that can cut off limbs or cause other injuries (Bailey et al., 2010; Borovnik, 2011). It is also much harder to grasp a rope when it is cold. Many seafarers have a vivid memory of their first deployment on ships, and they often cite weather as a factor that exacerbates the demanding, intense work. As mature, experienced men look back at their first journeys, they compare these challenges compassionately. When I had asked whether he liked his job, this established I-Kiribati seafarer, in the following extract, thought that it was very hard work. He reminisced about his first ship in winter, and how his shipmates explained to him that this type of hard work was 'normal', so he had to better get on with this job, no matter the hardship in winter. Yet, he was struggling in the cold:

> It's a very hard job. The weather is one thing.
> *The weather? Why? Because it is cold?*
> Yes. My first ship in New York... in wintertime! And we have to, you know this anchor, on the bar of the ship ... you know this anchor... Where you use to drop it, when you just [drop] it out further in the water. It's full of ice!
> *Full of ice?! You couldn't anchor it?*

Yeah! Stuck!

And what did you do?

And then the bosun: 'That's normal work of the seamen sometimes'. Ok, we have to go out and – take a big hammer and...! [he shows a strong hammering with his hands]

And you have to be very strong and push.

Strong and push. And all [of us] get overboard [to free up this anchor]. ... Yes. The weather! You are not get used to the weather. You are not adapted to it. That means you are not born and grown up with this cold weather. But you have to make it! You see, it is your job. Just what I always do. Some... have no strength. Some... they quit. You know? But it's very minor. It's very little that they quit.

It's not because of the weather that they would quit?

Yeah. Some of them they do. Also, because of, you know the psychology of how you live in a different atmosphere. [You are] always on the ship. Sometimes you get lonely, yes?

In comparison to the enjoyment expressed above by this Tuvaluan when his tanker had become stuck in ice in the Scandinavian Arctic, this man remembers how difficult and tough it was to launch an anchor in icy conditions. In his situation, it was gruelling work having to climb overboard, hammering into the ice to free it for the anchoring action. He then contemplated and explained the challenge of not only having to deal with an environmental atmosphere that is almost the opposite of Kiribati but also how the difficulty of dealing with draining weather situations was exacerbated by being on the ship for what seems like an endless time. This is why he had seen many of his friends quit their jobs. At this time, during the late 1980s and early 1990s, he and others had to spend 17 months working on ships. He felt stuck in more than one way when the ship was suspended in the ice in the wintery New York harbour. And yet, there is pride in the Kiribati culture. A man becomes *real* (a *te aomata*) in fulfilling his job as a 'seaman' and taking on hardship (Itaia, 1984). Hence, members of the Kiribati ship-crew remind each other to keep going and to look at the weather as something 'minor'.

A retired, elderly Tuvaluan seafarer described to me a time when he worked on a tugboat in the Malaysian area, where Tuvaluans and Malaysians worked well as a team. Then, almost instantly relating weather to stress, he also remembered the Canadian winter. I also attend here to expressions of pride in communal values, in this case the 'Tuvaluan-ness', a national pride wherein Tuvaluans are brought together in an image of a strong, resilient grouping:

[Tuvaluan seamen] are very happy people. They, I think, they really enjoy life also. But physically they were disadvantaged in the physical demands of the work. We happen to work on the tugboat, when I worked with the Malaysians. So, all the physical work... but it was because of our close association with them [Malaysians] that we, it was like a small brother... 'take this piece of iron from you', like that. And they were happy for us to do it, you know.

Yeah. How many Tuvaluans were on that boat when you...

There were four of us.

And you would help each other in that way?

Yeah, because it was the only way, you know to stand the stress on board such a high stress environment on the tugboat. The tugboat itself is, you work hours on the tugboat, it is a six [hours] on – six off [job]. ....
And then you get on and after a week or two then you really start feeling the physical stress.

Unusual hours, yeah.

But if you do a lot of anchor handling and removing, like that, and when the weather is not so good, I mean, or you're running the extra supply runs, because they are needed urgently in another place, you tend to feel the stress. Even mentally you bear, and that's why I say I'm quite happy to state the Tuvaluan guys are really resilient and we stand a lot of stress. Just imagine also the Tuvaluan who is taken immediately from here into mid-winter, and he boards his ship in something like northern Europe or Canada, in Vancouver. Just imagine that. But yet, after the first day or two, just a little bit of easy work amongst ourselves; keep away from the cold and the damp, a day or two, he will not like to do that. He wants to be out there where the action is, because he doesn't want people to regard him as, you know, running away from your work and letting all his comrades doing all the work. He wants to be fully engaged immediately, and he will feel guilty that he's being treated like a sissy. So, he's got to go straight up. But just imagine from a hot Tuvaluan... like this [looks around in the heat of Funafuti Island] in Tuvalu, and you go [from here] immediately in the middle of winter to Vancouver, get a ship there, and you start work immediately, it's not fun.

It is difficult to work in strong weather conditions, having to adjust to weather that is different from the tropical islands. Yet, it is important for a Tuvaluan to overcome stress and dislikes and to straighten up and adjust swiftly, and identify with overcoming obstacles and hardship, so that he is treated as a masculine, capable seafarer, ready for any conditions. Fajardo (2002: 180) argues that 'masculinization, particularly through the figure of the seaman, is attempted, manipulated, naturalized, and reinforced' through feminisation, and gender is also 'racialised and classed'. Although Fajardo refers to a Filipino context, this could be extended to other seafaring cultures. The above shows how important a masculine understanding of seafarers is, not only within the seafaring culture but combined with the Tuvaluan or I-Kiribati understanding of manhood as tough and resistant to quite challenging weather conditions. These quotes provide a glimpse into the embodied realities of global seafaring employment and the extent that they are dictated by the changing weather, seasons and climate that ships move through. They show the importance of cultural values, of pulling together as individuals who are part of a larger, communal entity. They also show the constraints of the seafaring jobs, where work has to be done within time pressures and in all kinds of weather. This kind of work stress is exaggerated when weather conditions intensify, for example in fog, when the visual decreases and seafarers have to use all their senses to move on, or with rain or snow, or when wind causes higher waves, or turbulence, and even tornado-type conditions.

Conclusion

Weather and ocean engage in different dances, where at times the ocean takes over – creating strong or flat waves and currents of different directions; at other times, the weather leads with an orchestra of sun, light, clouds, rain, cold and ice. When the ocean freezes, it 'suggests a *momentary* stopping point' (Fajardo, 2002: 177), which mobilises, once the weather allows it. Similarly, Vannini and Taggert (2014: 93) have observed icy-water materials as 'constantly changing' and intersecting. This changeable relationship seems to be becoming this or that randomly without specific starting or ending points. The fluid, mobile, watery ocean is journeyed through, while seafarers are moving with it, working on container ships, tankers or cruise ships in different shipboard sectors.

One may consider seafaring as an ongoing battle with extreme weather phenomena, such as typhoons or rough seas, where they have, over centuries, been encountering possible 'danger, disease, and death', as Urry (2014: 157) observed. Notwithstanding, I have shown in this chapter how weather, even the constant conditions such as heat and sun, or cold and icy waters, evoke bodily sensations that inform seafarers' awareness of their coming into daily contact with the sea. I focused on narratives of the everyday examples of their profession and their demanding work environments and how the nuances of weather that are not highlighted in logbooks as reportable affect seafarers profoundly. These everyday encounters are never predictable: the line of a travelling whale coincides with the circuit of a cruise ship; the stillness of thick ice in the New York harbour occurs just when a first-time I-Kiribati tries to launch the anchor; and his parents wanting to touch snowflakes reminds a Tuvaluan man of other impossibilities. Even small weather experiences can momentarily become miraculous and transformative, such as the wonder of snowflakes. These coincidences, and wonders, where seafarers touch or see *in* the weather (as in Ingold, 2007), are also formations of becoming-seafarer. In this chapter, I have extended becoming-seafarer to their communities. Deleuze and Guattaro (2011: 256) argue that 'only with other affects, with the affects of another body' (human or nonhuman), can we begin to know more about each other. Families are indirectly transformed and extended through seafarers' becoming-ocean/weathering encounters. Hence, weather-related processes on the deterritorialised sea may become disruptive, but they are not disconnecting processes. They instead produce and reproduce interconnections.

References

Alaska Shore Excursions (2019). Whale migration patterns throughout Alaska. Access online: https://alaskashoreexcursions.com/blog/whale-migration-patterns-throughout-alaska

Anderson, J and Peters, K (2014). 'A perfect and absolute blank': human geographies of water worlds. In: Anderson, J and Peters, K (Eds.), *Water Worlds: Human Geographies of the Ocean* (pp. 3–19). Ashgate, Farnham, Burlington.

Bailey, N, Ellis, N and Sampson, N (2010). *Safety and Perceptions of Risk: A Comparison between Respondent Perceptions and Recorded Accident Data.* SIRC, Cardiff.

Borovnik, M (2011). Occupational health and safety of merchant seafarers from Kiribati and Tuvalu. *Asia Pacific Viewpoint*, 52(3), 333–346.

Borovnik, M (2017). Night-time navigating. Moving a containers ship through darkness. *Transfers: Interdisciplinary Journal of Mobility Studies*, 7(3), 38–55.

Borovnik, M (2019). Endless, sleepless, floating journeys: the sea as workplace. In: Brown, M and Peters, K (Eds.), *Living with the Sea: Knowledge, Awareness and Action* (pp. 131–146), Taylor and Francis/Routledge, Abingdon.

Brown, M (2015). Seascapes. In: Brown, M and Humberstone, B (Eds.), *Seascapes: Shaped by the Sea* (pp. 13–26). Ashgate, Farnham.

Connolly, WE (2011). *A World of Becoming.* Duke University Press, Durham, London.

Deleuze, G and Guattari, F (2004 [2011]). *A Thousand Plateaus. Capitalism and Schizophrenia* (Trans. Brian Massumi). University of Minnesota Press, Minneapolis, London [originally 1987].

Fajardo, KB (2002). *Filipino Crosscurrents. Oceanographies of Seafaring, Masculinities, and Globalization.* University of Minnesota Press, Minneapolis, London.

George, R (2013). *Ninety Percent of Everything.* Henry Holt, New York.

Greenfieldboyce, N (2011). Steady as a whale? Humpbacks swim straight lines. *National Public Radio*, 22 April. Online access: https://www.npr.org/2011/04/22/135639670/steady-as-a-whale-humpbacks-swim-straight-lines

Hays, GC (2017). Ocean currents and marine life. *Current Biology*, 27(11), R470-R473.

Hull, J (1997). *On Sight and Insight: A Journey into the World of Blindness.* Oneworld Publications, Oxford.

Hutchins, E (1996). *Cognition in the Wild.* Massachusetts Institute of Technology Press, Cambridge, MA, London.

Ingold, T (2007). Earth, sky, wind, and weather. *Journal of the Royal Anthropological Institute (N.S.)*, 13(1), S19-S38.

Ingold, T (2011). *Being Alive: Essays on Movement, Knowledge and Description.* Routledge, London, New York.

Itaia, M (1984). Rebirth. Te Mauri, Te Raoi, ao te Tabomoa'. In: Talu, Sister Alaima et al. (Eds.), *Kiribati, Aspects of History* (pp. 121–128). Kiribati Ministry of Education Training and Culture, Tarawa.

Langewiesche, W (2005). *The Outlaw Sea. A World of Freedom, Chaos, and Crime.* North Point Press, New York.

Martin, C (2011). Fog-bound: aerial space and the elemental entanglements of body-with-world. *Environment and Planning D: Society and Space*, 29(3), 454–468.

Merleau-Ponty, M (1962). *Phenomenology of Perception.* Routledge and Paul Kegan, London.

Merleau-Ponty, M (2003). *Nature: Course Notes from the College de France.* Northwestern University Press, Evanston, IL.

Peters, K (2014). Drifting: towards mobilities at sea. *Transactions of the Institute of British Geographers*, 40, 262–272.

Raban, J (1999). *Passage to Juneau: A Sea and Its Meanings*. Picador, London.

Stanley, J (2016). *From Cabin 'Boys' to Captains – 250 Years of Women at Sea*. The History Press, Stroud.

Steinberg, Ph and Peters, K (2015). Wet ontologies, fluid spaces: giving depth to volume through oceanic thinking. *Environment and Planning D: Society and Space,* 33(2), 247–264.

Symes, C (2012). All at sea: an auto-ethnography of a slowed community, on a containership. *Annals of Leisure Research,* 15(1), 55–68.

Thrift, N (2008). *Non-Representational Theory. Space, Politics, Affect.* Routledge, Abingdon, NY.

Urry, J (2014). *Offshoring.* Polity, Cambridge.

Vannini, P (2013). Slowness and deceleration. In: Adey, P, Bissell, D, Hannam, K, Merriman, P and Sheller, M (Eds.), *The Routledge Handbook of Mobilities* (pp. 116–124). Routledge, London.

Vannini, P and Taggart, J (2014). The day we drove on the ocean (and lived to tell the tale about it); of deltas, ice roads, waterscapes and other meshworks. In: Anderson, J and Peters, K (Eds.), *Water Worlds: Human Geographies of the Ocean* (pp. 89–102). Ashgate, Farnham, Burlington.

Vannini, P, Waskul, D, Gottschalk, S and Ellis-Newstead, T (2011). Making sense of the weather: dwelling and weathering on Canada's Rain Coast. *Space and Culture,* 15(4), 361–380.

8   Snow matters

From romantic background to creative playground in alpine tourist practices

Martin Trandberg Jensen and Szilvia Gyimóthy

I mourn the loss of winter snow with an existential grief and a feeling that life will no longer be the same.

(Schmidt, 2020)

Introduction

This chapter explores the performative potential of snow in tourism, by drawing attention to its material and non-representational significance for tourism practices. Snow plays a central role in the cultural imagination of the Alps, and even today, snow features as the *sine qua non* attribute of high-altitude mountain destinations. Hitherto, researchers mostly engaged with snow-clad landscapes as a backstage as well as with symbolic and representational qualities of snow. Despite snow being an essential element for tourism in the Alps, little work has been done to understand how snow shapes the creation of affective atmospheres and sensations. More precisely, our focus is directed at tourists' embodied practices and engagement with snow, which permits the assessment of its qualitative properties allowing (or resisting) movement and temporary dwelling in snow-covered terrain. Drawing on discussions on the politics of snow as well as on ethnographic observations of 'not-snowborne' tourists tumbling in the snow, the chapter argues for the significance of unintended, creative and playful affordances of this weather phenomenon. Mundane playful practices with/in snow may improve one's motricity and balance, but they also accentuate new appropriations and design potentials for this elusive substance, that are not well-acknowledged by planners and product developers. The chapter concludes with a discussion on micro-design issues emerging from the analysis and offers future perspectives for snow destinations.

The symbolism of snow in cultural representations: magical, pure and perilous

Snow White. Miss Smilla's Feeling for Snow. The Ice Queen. Frost. European imagination has been preoccupied with snow since medieval times. Contemporary popular cultural representations keep on recirculating coherent symbolic associations from classic fiction and visual arts about snow and its dreamlike character. Snow transforms a familiar landscape into an other-wordly, sparkling utopia in which glittering snow crystals and icicles light up the *winter wonderland.* Crooning along with Bing Crosby throughout the holiday season, we are desperately longing for a *white* Christmas and sleigh bells in the snow. Indeed, the materiality of snow is magical. In children's stories and songs, snow is benign and wraps houses, parks and streets with a soft, fluffy, temporary duvet. Snow alters our auditory and kinaesthetic senses: it muffles noises, slows our movement or impedes traffic. Freshly fallen snow gives a sense of 'untouchedness' to places. The white colour signifies *immaculata* [spotless] qualities, imbued with associations of purity, serenity and innocence (Hansson and Norberg, 2009). However, snow is by no means an innocent substance; it can become bleak and hostile in large quantities. Even today, when it comes as blizzards and avalanches, snow represents a natural superpower that is beyond the control of humans. Not surprisingly then, cultural imagination often depicts the empire of eternal snow as a realm beyond the human world. It is an exotic and dangerous place that is not meant to be inhabited by humans, but by fabled figures and creatures such as snowmen, elves and fairies. In Hindu mythology, snow-clad mountain tops are associated with the spaces of the divine, and are equally inaccessible for mortals (Schneider, 2009).

These ambivalent qualities are often transferred to the imaginaries of conquering (by moving into and up) high mountains and polar areas. Romantic literature and art from the 18th century depict the Alps as an unpolluted and unspoilt territory far from civilization, which can be admired and gazed upon from a distance. However, the exploration of the realm of snow and ice can be a life-risking endeavour, and by no means a dreamy 'walk in the woods'. In most literary representations, the High Alps are described as harsh and frightening, only to be scaled by courageous, strong men. For centuries, these landscapes represented the ultimate frontier and posed a challenge for privileged male adventurers. Explorers daring to venture into the white wilderness were driven by the belief that 'such a thing as a moral triumph over the snows was possible' (Spufford, 2010: 269). In Francis Spufford's view, the snow becomes the enemy, about which he exclusively talks in terms of conflict, conquest and struggle. The adversary, masculinist discourse of 'Man' vs. Wild is still prevalent in contemporary popular culture, and frequently reproduced in feature films and reality shows (such as the *Alone* franchise) and tourist promotion.

While the individualistic explorer quests of the 19th century are clearly inspired by both romanticism and colonial imperialism, early 20th-century alpinists' view of the ultimate snow frontier is more progressive and technocratic. For instance, analysing the Deutsches Alpvereins infrastructural developments in the Easter Alps between 1885 and 1915, Anderson (2012) documents the emergence of the modern gaze on the Alps. 'The region of once feared snow' is gradually turned into a 'Bürgerlich' cultural resource through the construction of roads, paths and huts. During the dawning tourist interest in mountain space, Alpinist associations saw it as their moral task to transform this inaccessible landscape into a comfortable pleasure zone for the wider public (ibid.). Throughout most of the 20th century, the leisure-centred logic of urban elites defined alpine touristification (Nahrath and Stock, 2012) and also affected individual practices beyond architectural interventions. This could entail the emergence and consolidation of new fashion styles for mountain visitors as well as the adoption of 'appropriate' ways of walking and moving in the snow (often involving material equipment, such as special snowshoes, skis, sledges and snowmobiles).

In the postmodern era, snow-clad landscapes are further tamed and commoditized for leisure consumption. Skiable domains stretching across several valleys have turned uninhabitable corners of the Alps into outdoor playgrounds. Cable cars lift visitors who possess no extraordinary physical attributes to peaks above 3,000m, where comfortable shopping and entertainment facilities await them. After the turn of the millennium, alpine experiencescapes increasingly draw on technologically advanced snow substitutes through the use of artificial snow or virtual reality. However, although alpine nature is increasingly anthropogenic and groomed, the commercial narratives of skiing destinations and high-altitude attractions still take their inspiration from other-worldly discourses. It is equally remarkable, that the mythical realm of snow is often intensified through vivid sensory descriptions. For instance, the portrayal of the first-hand glacier experience on Jungfraujoch – Top of Europe is conveyed through inimitable visual, tactile, kinaesthetic and respiratory impressions:

> Icy air *streams across* your face, snow *crunches* under your feet, and the view nearly *takes your breath away*: on one side the view of the Mittelland up to the Vosges, on the other the Aletsch glacier, bordered by four thousand metre peaks. Standing on the Jungfraujoch 3,454 metres above sea level, *you can feel it* with your first step: *this is a different world*. It's one you have to experience.
>
> (Jungfraujoch, 2017)

Along cultural interpretations, these sensuous perspectives beg for exploring how snow matters from a non-representational perspective. Snow is a visceral and embodied material deserving a more-than-interpretative analysis. We argue that non-representational theories and sensuous approaches to mobilities research offer a novel way of engaging with snow and its animate materiality. Our claim is investigated through a reflective analysis of the poetics and politics of snow, through methodological considerations of doing fieldwork in the snow, and by drawing upon empirical snapshots of tourist practices in the snow at Swiss high-altitude visitor sites. In conclusion, we synthesize these three streams of discussion and provide a few suggestions for future studies on weather, snow and contemporary tourism mobilities.

A non-representational glance at snow in tourism

In an attempt to go beyond and nuance the grand narratives and symbolic and historical accounts presented in the preceding section, we speculate on the value of adapting non-representational theories (Anderson and Harrison, 2012; Lorimer, 2005; Thrift, 2008) to the study of snow and tourism mobilities. Set in the context of tourism research, the adoption of non-representational theory is not entirely new, and our theoretical approach extends and contributes to other recent accounts in areas of research such as tourism mobilities (Hannam et al., 2014), alternative sport activities (Thorpe and Rinehart, 2010), sensuous tourism (Edensor and Falconer, 2012; Jensen et al., 2015) and epistemological discussions about tourism knowledge-creation (Xiao et al., 2013). To avoid early misconceptions, non-representational work cannot be characterized as *anti*-representational, but is better understood as 'post-representational' (Vannini et al., 2011) or 'more-than-representational' (Lorimer, 2005) in its attempt to

> take[s] representation seriously; representation not as a code to be broken or as an illusion to be dispelled rather representations are apprehended as performative in themselves; as doings. The point here is to redirect attention from the posited meaning towards the material compositions and conduct of representations.
>
> (Dewsbury et al., 2002: 438)

Accordingly, our account breaks with preoccupations with 'representationalism or discursive idealism' (ibid: 438), where the focus is on what things symbolise, what they denote and connote, what codes they inform, what values they defer/refer to (Vannini, n.d.). Non-representational theory represents multiple schools of thought that focus on the significance of embodied practices in space, and indeed, recent non-representational theory–inspired research has sought to examine the complex relationships between weather, humans and physical landscapes. As Rantala et al. (2011) suggest, weather is much more than a mere medium between humans and the environment; it itself possesses powerful agency. From their practice-oriented study of wilderness guiding, they identify three types of weather-related practices: *anticipating*, *coping* and *discursive*. Most relevant for this chapter, they explore how a guide's ability to 'read' the weather is an example of an embodied gut-feeling and lifetime skill. Here, the guide is not reduced to a body-less and educated mind whose intellect has mastered how to interpret weather forecasts. Quite the contrary, a large proportion of the weatherly anticipation, they suggest, emerges through the guide's non-representational, pre-cognitive and multisensuous impressions in specific situations. Similarly, we downplay practices-in-the-snow as an expression of deep identity traits, routines and cognitive thought processes, and favour an analytical focus on the *networked* and largely *intuitive* and *pre-cognitive* processes through which a range of actors, including snow, clothes, material infrastructures, atmospheres, sensations and humans, work together to inform the experience and meaning-making of alpine tourist experiences.

Interestingly, while the majority of alpine tourist practices are set around relatively disciplined and staged socio-material performances, such as skiing and snowboarding (Edensor and Richards, 2007) or alpine trekking (Baur et al., 2012), our research focuses on relatively undisciplined 'bodies-in-the-snow'. For this purpose, non-representational theory's focus on capturing the 'on-flow' of everyday life (Thorpe and Rinehart, 2010) is particularly useful as it allows us to delve into how lived sensuous life, imagination and even play are integral parts of becoming a snow-familiarized alpine tourist. Our non-representational approach exemplifies how a number of alpine tourist practices are informed by both affective and sensuous orientations set in snow-clad space. The ways in which tourists make sense of snow is informed by kinaesthesia, audition and tactility. For international tourists visiting the alpine regions of Switzerland for the first time, their curiosity and engagement with snow is marked by a multifaceted, playful, embodied, experimental and intuitive disposition. For this reason, non-representational theories are useful to elucidate 'the contingent, the ephemeral, the vague, fugitive eventfulness of spatial experiences' (Jensen and Lanng, 2016: 65). This particular focus on spatial experiences also means that we acknowledge the agential role of technologies, objects and the physical landscapes and how they are implicated in what the body is and what the body is likely to do at particular moments (Macpherson, 2010). Importantly, when we talk about the materiality of snow, we do not refer to any type of fixed and non-permeable substance or an inert icy surface upon which human action unfolds. Rather, the various piles and concentrations of snow are looked upon as 'meshworks' (Ingold, 2011), constantly undergoing change, rupture and movement. The snow that we study is thus, similar to 'ice roads' (Vannini and Taggart, 2014), a mutating, temporary and ephemeral landscape emerging from the changing relations between humans, weather, animal life and industrial activities.

The poetics of snow

As mentioned earlier, a core trait of non-representational thinking is the significance of materiality in everyday life. Such material interest is inspired by the recent social sciences around critical animism, posthumanism and alien phenomenology (Bennett, 2010; Bogost, 2012; Vannini and Taggart, 2014). Through insisting on a vibrant materiality, non-representational scholars critique conceptual approaches that reduce materials to dead, passive objects, and point out that they are unable to more fully analyze their situated, qualitative and transformational properties and effects (Ingold, 2012).

To explore the intricate details of snow, the notion of 'affordance', originally coined by Gibson (2015[1979]), is pertinent. Affordance refers to the practice potentials of specific objects or environments, and should not be understood as innate qualities of objects but rather as mediated effects emerging from the interactions between humans and environments: 'The enrolment of the human body into designed systems, sites and infrastructures creates complex assemblages where materialities are not just external to the human, but rather permeable' (Jensen, 2016: 593).

Think back, for a moment, on childhood practices of preparing snowballs. Through multiple winter seasons, many of us have accumulated embodied experiences of 'compressing snow with hands', and we know how temperature conditions the stickiness and granular features of the ideal snowball. We might recall that the texture of dry snow is inadequate to make good snowballs. The fresh ice crystals, it seems, do not stick too well. They are too porous to compress and harden, and the crystals fall between your fingers. Conversely, wet snow provides a perfect malleable substance that can be easily amassed, compressed and formed. We may even recall learning skills, such as pouring water onto the snow to harden it quicker, to engineer the perfect 'lethal' snowball for playful snow fights. With the advent of climate change–induced warm and wet winters, such embodied knowledge may become obsolete – along with the vocabulary that describes these sensations. As Gudrun Marie Schmidt put it so eloquently in her essay on the 'last winter kids':

> Frost, icicles, snowstorm...Who will use these words when what they describe is disappearing? Who should talk about the creaking feeling of the frosts under the soles? About the icy blue shadow in the footsteps of snow? About the flash of diamonds in the powder snow that the sun calls forth in all white?
>
> (Schmidt, 2020, our translation)

Accordingly, when thinking about the affordances of snow we have to understand the qualitative properties of different types of snow and the practices they afford. Despite snow being an inedible substance, the figurative terms depicting various snow granularities are often inspired by food-related metaphors. *Corn*, for example, is a commonly used expression by skiers to describe good spring snow characterized by its coarse, granular and 'wet' qualities. Another popular snow type is *champagne powder*, which refers to very smooth and dry snow, particularly good for skiing. In opposition, *ice* is a compressed form of snow that can be smooth, slippery and extremely difficult to move on as a skier or alpine tourist. And these popular terms only represent the snowboarding and skiing cultures. According to Magga (2006), up to 1,000 lexemes with connections to snow, ice, freezing and melting can be found within the North Saami language, pointing to the central role of snow and ice conditions of Arctic living. Consequently, from this complex understanding of snow, we may start to imagine how for a skier, different coping skills and materials, experiences, intuitions and 'snow feels' are central in skiing performance (and similarly for the local Saami who makes a living from the opportunities allowed for by the icy conditions).

The politics of snow

Snow is not merely a natural substance or cultural object of scrutiny but a complex phenomenon with fundamental economic implications for local communities. With the rise of ski tourism, snow has become the central 'currency' in alpine societies, which are highly dependent on increasingly unpredictable winter precipitation. To avoid the obliteration of entire resort industries and economic systems, snow is turned into an engineered material: *artificial snow.* Compared to natural snow, which originates in molecules of water vapor, floating high in the atmosphere, in temperatures that hover around freezing point or below, human-made snow is compressed air shot through a nucleator nozzle that mixes water droplets with cold air to produce frozen snow grains (circular grains with no resemblance to the hexagonal structure of the natural snowflake). To hold together the natural-and-artificial snowscapes of popular alpine skiing resorts, a large and complex socio-technical system comprising alpine workforce, service providers, snow canons, ski lifts, snow fences and snowploughs are necessary. In coordinating this system, the manufacture, control and (re)distribution of snow caters for specific alpine leisure activities. As these systems are often dominated by entrepreneurial and neoliberal rationales, we see the contours of a *politics of snow*, which is changing the nature of snowscapes as we have traditionally perceived them.

From broadening out our material understanding of snow, and underlining the role that people play in networking it into meaning, it is possible to imagine how being an international tourist, initially unaccustomed to snow in any of its many forms in an alpine setting, is a process of becoming an 'able-bodied' snow tourist: the tourist must learn how to navigate snow-clad landscapes, develop distinct material practices to afford movement on snow, become familiar with snow through sensuous orientations and experiments, playfully use snow as props for staging activities, and slowly attune to the properties, propensities, problems and potentials of snow.

While such descriptions may appear banal, attuning our critical attention towards the affordances of snow allows us to address a range of interesting, but neglected, questions: how do certain materialities (such as snow) feel?

How is it experienced? What are the practical consequences of this? How is snow re-appropriated, circumscribed and practised differently by tourists, and with what kinds of different end goals? In addition, since alpine resorts are always partially staged 'from above' (Jensen, 2013) through design intentions (infrastructures, building principles, regulations, technologies and machines), who/what are specific snowscapes catering for? Who is marginalized? How can the material environment be changed to work with different effects, such as functional, ethical, environmental or social? To further explore these questions, we now draw upon a number of ethnographic examples from fieldwork in Switzerland.

Doing fieldwork in the snow

Ethnographic examples are derived from a ten-day fieldwork conducted by one of the authors during May 2017, which aimed to explore overseas visitor experiences in Switzerland and the textural transformations of Switzerland's increasingly cosmopolitan touristscapes (Gyimóthy, 2018). Owing to warming temperatures and dwindling snow conditions, alpine railway and cable car operators are increasingly targeting a wider public of mountain novices, many arriving from emerging Asian and Middle Eastern markets. The initial intention was to study tourist performances conditioned by contemporary non-Western imagination of the High Alps (such as Bollywood movies). During the scoping of the project, the three most-frequented high-altitude locations of international mass tourism were selected, namely Engelberg near Lucerne (Mount Titlis), Interlaken (Jungfraujoch – Top of Europe) and Gstaad (Glacier3000), with Jungfraujoch topping the list with over a million visitors every year (The Local, 2019). These destinations' iconicity was not only elevated by James Bond films but also several Asian screen productions, which earned them a mandatory spot on Indian tourists' "Yatra" [journey] to Switzerland. Hence, it was anticipated that new tourist performances will reflect Asian popular cultural representations and romantic imaginaries distinct from the Eurocentric ideals discussed above.

From the outset, it was not difficult to distinguish between visitors familiar with alpine environments and mountain neophytes only on the basis of walking styles, body positions and balancing techniques. This distinction was equally reflected in appearances, which turned out to be far more creative and eclectic than one could expect from destination promotional material. Instead of adhering to the imperative of layered outdoor clothing and technical mountain footwear, visitors would typically wear chic fashion garments appropriate for city excursions (or colourful saris and kurtas) as well as sneakers, sandals or flip-flops. Group travellers often sported the tour operator's identical, bright-blue sunglasses and preferred to stay close together throughout the entire trip. It was not unusual to see grown-up and elderly Asian tourists cheerfully frolic in the snow on the top stations or generously share both spicy snacks and sunscreen with other tourists – inclusive of the researcher. However, instead of imposing a cultural interpretation on alternative looks and walks (as signifying practice from a particular region of the world), we chose to see them as intuitive engagements with unfamiliar spatialities, materialities and weather phenomena.

Hence, beyond the planned and intentional data collection, the rich audio-visual material from the field trip revealed unexpected insights into the intuitive doings of visitors inexperienced with snow and high altitudes – standing in stark contrast to those of skiers and mountaineers. Tourist practices were often compromised by extreme or rapidly changing weather phenomena and ambient conditions. Below zero temperatures, lower air pressure, striking sunlight, icy winds and dense fog had a direct impact on the accomplishment of a visit, let alone on conducting participant observation and engagement with fellow visitors. Within such conditions, adaptable and heuristic tactics proved to be more effective than strictly scheduled data collection. Accordingly, during the course of the fieldwork, a distinct form of methodical opportunism was developed that revolved around exploiting trivial and inoffensive bonding opportunities afforded by public spaces. Small acts of generosity, such as taking pictures, joining in group photos, offering sunscreen, accepting invitations for shared food consumption and being stand-in tour guides, provided an intimate platform for meeting and engaging with other visitors on equal grounds.

In the following sections, we approach these incidental fieldwork insights by capturing non-snowborne visitors' *feeling for the snow*, which allows us to reflect on the impact of a radical 'element shift' of being-and-moving-in-the-snow. The empirical illustrations will be analyzed along four

Figure 8.1 Getting ready for the high alpine experience.
Source: S. Gyimóthy.

mobility-related themes: tumbling and learning to move, sensuous playing with/in snow, mediatized performances and being insulated from snow in interior spaces.

Tumbling in the snow: becoming snowborne

Let us turn our attention to the very first corporeal encounters with height, sun and snow and sensitise with extreme ambient conditions. Taking the cable car to 3,000 meters in five minutes makes the stomach sink, but also other atmospheric effects put the body under considerable stress. The air pressure drops by 30%, ultraviolet exposure (on sunny days) increases by 40% and the albedo effect (diffuse reflection of solar radiation) rises to the maximum. The rapid change in altitude may trigger acute nausea. The blinding reflections of the sun may cause headache, even sunstroke, despite the chilly temperatures. These physiological responses provoked fuzzy emotions among first-time visitors, whose anxiety and awe were reflected by unarticulated vocal reactions. Interestingly, individuals' pre-cognitive reactions were almost contagious, especially if they were drawn together in a confined physical space, such as a cable car carriage. Quite often, excitement and nervousness could lead to intuitive collective performances. For instance, when the cable car set off, Hindi groups would suddenly break out in chanting, praying to arrive safely to the top. When gale and snowfall forced the operators of Glacier3000 to temporarily halt all chairlift operations, most members of two bus groups took the opportunity for a nap on any horizontal surface. Men and women of all ages casually lied down on the restaurant floor (or even along the window sills), emulating mooring practices from waiting rooms in other transit spaces. Such incidents excellently illustrate how high alpine experiences are afforded by networked entanglements of mobility infrastructures (cable car), atmospheric conditions, bodily sensations and social group dynamics.

Upon arriving to the top of Jungfraujoch, Glacier 3000 and Mount Titlis, visitors had the opportunity to wander outside on the glacier to get a hands-on experience of snow. Exiting the station building, visitors' movements reflected discomfort and insecurity; their cramped bodily postures reminded us of Bambi taking his first steps on ice. Snow distorted their kinaesthetic senses and proprioception, and sandals – which are perfectly suited to keeping a solid stance on the sidewalk – turned into a rather uncontrollable footwear on a slippery substance. Snow requires a whole different way of walking and other bodily competences – and visitors employed vastly different coping techniques to acquire those skills. Younger adults would rush out of the cable car to be the first to touch snow, and giggle when slipping uncontrollably on the icy surface. More cautious tourists would make small strides without lifting their feet. They glided one foot carefully after the other, often anxiously holding on to handrails or their travel partners. Yet others would develop creative tactics to stay upright. For instance, three mid-aged women sat down to put on rubber bands on their sneakers, making sure to plant their feet firmly in the snow.

Other uncommon atmospheric phenomena equally afforded snow neophytes' innovative coping techniques in the snow. In bright sun, they would cover their heads with their jackets or sweaters to avoid being exposed to the reflections and heat of sun from white surfaces. During days with strong chill factor (strong winds, high humidity and/or snowfall), elegant elderly ladies would complement their fashionable attires with warm blankets or balaclava headwear. They would rather give up being mountain chic than not going out in the snow. These creative solutions indicate that the 'element shift' not only requires the mastering of new bodily skills and muscular control but also the incorporation of a vast range of material objects in engaging with extreme weather. These illustrations resonate Rantala et al.'s (2011) notion of weather phenomena being an agency of creative corporeal learning. Becoming snowborne is paved with sensuous surprises, bodily discomfort and frustrations. However, as the next empirical section suggests, mastering moving in the snow is not a goal-oriented endeavour, but a ludic and explorative process.

Playing in the snow: sensuous delights

The skills that visitors learned were often acquired in playful ways. The unfamiliar materiality of snow was explored intuitively – as children might do so – using not only the haptic sense but also taste. Visitors would often take a bit of fresh snow in their hands and attempt to mould it into a shape

Figure 8.2 Improvising anti-slippery soles with rubber bands.
Source: S. Gyimóthy.

with their fingers. They would lick it or take a bite of a snowball to experience its chilly and elusive texture. Two children from an Indian family took up a handful of snow and put it inside their gloves, inviting their parents to do the same, while exclaiming in delight, 'Mummy, feel, it is hot and cold at the same time!' Gaining a (gut-)feeling for snow leads through such intense multi-sensuous engagements, that people on higher latitudes would instinctively acquire already in childhood. These universal tactile probings resonate well with Schmidt's bittersweet memories of frosty winters:

> I remember my sister's frozen hands after completed newspaper route. The sound of the skate scraping along the ice as we ran. The feeling of gravel in a snowball. The metal taste of the icicle, the melting water running down a playful tongue. The weight of the mittens after a snowball fight.
>
> (Schmidt, 2020, own translation)

In many ways, playing in the snow is a return to innocent embodied practices, and being a tourist in an exotic location exempted many adults refraining from being childish and irresponsible. On all three field trip locations, grown-up men and women rediscovered immature ways to open themselves out to the affordances of snow by making snowballs, snowmen or lying down to make 'snow angels'. People would spend less than five minutes on average on the panorama terraces, but they became completely oblivious of time while frolicking on the small sledging hill at Glacier3000 or while snowtubing on Mount Titlis. It was not unusual to see tour guides desperately herding and calling for their groups to return to the cable car. These venues offered a free-roaming, yet safe ground, where non-snowborne visitors could play their way to improving their motor skills for a more confident – and perhaps more graceful – movement in the snow. Yet, playful mobility practices in high altitudes and unpredictable weather conditions require vigilance. For instance, chairlift operators took extra care explaining how to advance to get on (and off) the chairlift, and seemed openly frustrated when groups of six attempted to mount a chair intended for a maximum of four people. At other occasions, undisciplined visitors were observed crossing beyond security lines (cords) to climb smaller rocks or wander a few meters into ungroomed (crevassed) terrain. Immediately, they would sink into knee-deep snow and would retract from exploring the perilous glacier on their own.

Mediatized performances: popular culture-induced romantic and sportive practices

Once becoming snowborne, visitors appropriated snow and snowy landscapes as a background for taking pictures. Apart from the most common photographing practices, we also noted remarkable romantic rituals, many

Figure 8.3 Touching snow for the first time.
Source: S. Gyimóthy.

of these paraphrasing scenes from iconic Bollywood films. Honeymoon couples would re-enact their marriage proposal, with the husband kneeling affectionately to ask for the hand of his wife, while having a third person throwing snow on them. On Jungfraujoch, a young woman spent an hour on the Aletsch Glacier, taking selfies in a white wedding dress. Seen from this perspective, snow remained a symbolic prop in a romantic spectacle, which stands in stark contrast to representations of the Alps as being spaces of adventure and conquest. Non-Western mediatized performances are turning Swiss mountains into a utopian landscape of love (Dwyer, 2002), which further enriches the more-than-representational texture of the Alps.

Besides romantic staging, other conspicuous performances included the mimicking of Alpinists or skiers, which required particular material objects. For instance, when noticing the skis carefully stored in racks at Mt Titlis, Indian tourists would pick a pair of skis to strike a pose with bent knees for the camera. Most of the time, they stepped on the skis with the tail in front, and involuntarily entertained fellow visitors with creative postures to regain control over their slippery equipment. After the successful photo session, skis were left helter-skelter on the ground, until other skiers would occasionally demand the wannabe-adventurers to put them back. Such unruly practices are disruptive from both design and discursive perspectives. First, they challenged the neat ordering of skiers' mooring systems, by temporarily appropriating the property of others to tumble in the snow. Second, they performed a new, hybrid conduct of representations: signifying both mass tourism and sport tourism performances at once. In such incidents, it becomes clear how alpine skiing resorts, as much as we associate

Figure 8.4 Posing with skis at Mt Titlis.
Source: S. Gyimóthy.

Figure 8.5 Tumbling in the snow.
Source: S. Gyimóthy.

them with relatively stable bundles of performances and operations, are anarchic spaces filled with heterogeneous practices. Being sensitive to the type and scope of such ordinary and easily neglectable performativities means 'tuning-in to the event-ness of the world' (Vannini, 2015, p. 321), and in so doing, coming to understand the open-ended and unscripted dimensions of alpine resorts. From such a more-than-representational perspective, any clear-cut definition of skiing resort performances is delineated, for as demonstrated, the subjects, cultures and performances that emerge in such contexts: '...come less from their place in a structuring symbolic order, and more from their enactments in contingent practical contexts' (Anderson and Harrison, 2012: 7).

Being insulated and disembodied from snow: indoors snow sensations

Although high alpine attractions enable intuitive, exploratory and disruptive engagements with the snow, most of the visitor experience occurs in regulated and ordered spaces insulated from snow. For instance, the last part of the train ride to Jungfraujoch takes place in a tunnel. Instead of direct exposure, visitors are presented with disembodied experiences of the Aletsch Glacier. In the train, a short informative video tells about the history and construction of the line, which is heavily supplemented with facts, statistics and security announcements. Adverts for local retail and food stores prime their potential customers for consumption, with reassuring emphasis on the high-altitude commercial arena. To give a sense of the void and wilderness above 3,000 metres, the train stops for five minutes at the Eismeer station. Still situated in the tunnel, Eismeer [Sea of Ice] features a panorama window to take pictures and to gaze at the glacier, safely shielded from the icy air or precipitation outside. In short, the journey to the Top of Europe station is anything but unmediated.

In a similar vein, the final station is also designed as a protective, artificial environment, with extravagant consumption opportunities *inside* the glacier, such as Europe's highest chocolate factory, a Swiss watch shop as well as panoramic restaurants. In cases of bad weather, visitors may stroll through the 250-meter-long Alpine Sensation Exhibition and the Ice Palace Cave, including an immersive glacier experience with multiplex screens, surround sounds and wind machines imitating the hostile beauty of the alpine environment. The Top of Europe is aligned with the spatial design of contrived shopping malls around the world, where indoor ski slopes and ice bars make the realm of snow and ice accessible for anyone. As a cosmopolitan lifestyle arena, it presents no sensory discomfort or bodily challenges to visitors; temperatures are kept on ambient levels and there are elevators and/or stairs connecting the different levels/realms of consumption. Not even on the outside panorama platform is it possible to feel the heavily promoted snow crunch under one's feet. Tourists' 'dwelling' in the High Alps becomes a disembodied experience, ordered by commercial logic and calculated sensory effects. The insulated places of Eismeer and the Alpine Sensation Exhibition offer smoother, easier and more familiar ways to move around, but simultaneously render snow into an inanimate and harmless substance that is only meant to be gazed upon. These alpine experiencescapes are engineered through deliberate design intentions and staged 'from above' (Jensen, 2013) by commercial stakeholders. In these contexts, the scope of alpine experiences is expanded further, and while the alpine tourist may be insulated from the actual snow, new configurations between the tourist and the alpine landscape are created.

Discussion...snow what?

The empirical discussion above demonstrates the refreshingly diverse ways in which non-snowborne visitors appropriated, circumscribed and practised snow-clad landscapes. Through light-hearted slipping and tumbling, through their sensory engagements as well as through improvised material practices and cultural performances, they learnt *how to be fluent* in moving in a new environment. Their creative and intuitive approaches to the acquisition of new mobility skills present new potential avenues with which to rethink the streamlined, artificial experiencescapes of snow destinations. High alpine attractions have created accessible and safe snowscapes but at the same time are imbued with a specific design infrastructure and particular technological apparatus that do not solicit playful explorations. This chapter therefore allows us to point at a number of interesting issues related to the snow mobilities design of mountain resorts. First, they are designed and managed to afford specific ludic practices that are primarily conditioned by modern Alpinism and commercial outdoor logic. These demarcate 'proper' ways of playing and moving in the snow, such as skiing and snowboarding, to fulfil visitor objectives of bodily mastery, quest and achievement. On the other hand, more sensory, improvisational and ludic practices such as rolling in snow or building with it are marginalized.

Conclusion

This chapter offers two key insights to the discussion of weather mobilities. First, we have taken a nuanced look at the dominant symbolic features of cultural representations of snow and snow-clad landscapes. By taking a non-representational approach to study a number of alpine snow performances, we have learned that snow, as ordinary as it appears, is heterogeneous and networked into significance through numerous interpretations, sensations and intuitive play. Future research should be aware of the complexity of this elusive snowy material before too easily reducing it to a symbolic token or instrumental material in tourist experiences. Second, through this approach, we have illustrated a number of ordinary, neglected and non-representational spaces within alpine resorts: the ephemeral, intuitively created and playful snowscapes. These snowscapes emerge in-between the existing dominant leisure activities (skiing, trekking, snowboarding) and outside the disciplining service system (restaurants, ski lifts, souvenir shops, alpine museums). As temporary socio-material snows-capes, these get-togethers are guided by intuitive and playful engagements with snow, where banal snowscapes are turned into mock battlegrounds, snowball-throwing arenas, snow-gliding corridors and simply unusual material grounds for bodily experiments and play, similar to the sandpits of children's playgrounds. With this analogy in mind, we can rethink the staging of alpine tourist regions: what if snow was more than the 'latent symbolic feature' of the alpine tourist landscape or the instrumental ground condition for skiing? What visions, development potentials or innovations may emerge if we imagine the resort as a playground? How can we think, plan and design the micro-materialities of alpine resorts differently to more thoughtfully afford heterogeneous and cross-cultural snow practices? These questions emanate from our situational fieldwork and we voice them with the hope for future research to respond to them through creative and critical fieldwork in the snow.

References

Anderson, BM (2012). The construction of an alpine landscape: Building, representing and affecting the Eastern Alps, c. 1885–1914. *Journal of Cultural Geography*, 29(2), 155–183.

Anderson, B and Harrison, P (2012). *Taking-place: Non-representational Theories and Geography*. Farnham: Ashgate Publishing Limited.

Baur, M, Kablan, E, Kasüske, D, Klauditz, A, Nordhorn, C and Zilker, A (2012). Trekking in the Alps? Spaces in trekking tourism from the perspective of Europe. *Journal of Tourism: An International Research Journal on Travel and Tourism*, XIII(2), 85–104.

Bennett, J (2010). *Vibrant Matter: A Political Ecology of Things*. Durham: Duke University Press.

Bogost, I (2012). *Alien Phenomenology, Or, What It's Like to Be a Thing*. Minneapolis: University of Minnesota Press.

Dewsbury, JD, Harrison, P, Rose, M and Wylie, J (2002). Enacting geographies. *Geoforum*, 33(4), 437–440.

Dwyer, R. (2002). Landschaft der Liebe. Die indischen Mittelschichten, die romantische Liebe und das Konsumdenken. In A. Schneider (ed.) *Bollywood. Das Indische Kino und die Schweiz*, Zürich, Hochschule für Gestaltung und Kunst, pp. 97–105.

Edensor, T and Falconer, E (2012). Sensuous geographies of tourism. In J Wilson (ed.), *New Perspectives in Tourism Geographies* (pp. 74–81). New York: Routledge.

Edensor, T and Richards, S (2007). Snowboarders vs skiers: Contested choreographies of the slopes. *Leisure Studies*, 26(1), 97–114.

Gibson, JJ (2015). *The Ecological Approach to Visual Perception: Classic Edition*. 1st edition. New York: Psychology Press.

Gyimóthy, S. (2018). Transformations in destination texture: Curry and Bollywood romance in the Swiss Alps. *Tourist Studies*, 18(3), 292–314.

Hansson, H and Norberg, C (2009). *Cold Matters: Cultural Perceptions of Snow, Ice and Cold*. Umeå: Umeå University and the Royal Skyttean Society.

Hannam, K, Butler, G and Paris, CM (2014). Developments and key issues in tourism mobilities. *Annals of Tourism Research*, 44, 171–185.

Ingold, T (2012). Toward an ecology of materials. *Annual Review of Anthropology*, 41, 427–442.

Ingold, T (2011). *Being Alive: Essays on Movement, Knowledge and Description*. New York: Taylor and Francis.

Jensen, MT, Scarles, C and Cohen, S (2015). A multisensory phenomenology of interrail mobilities. *Annals of Tourism Research*, 53, 61–76.

Jensen, OB and Lanng, DB (2016). *Mobilities Design: Urban Designs for Mobile Situations*. New York: Routledge.

Jensen, OB (2016). Of 'other' materialities: Why (mobilities) design is central to the future of mobilities research. *Mobilities*, 11(4), 587–597.

Jensen, OB (2013). *Staging Mobilities*. New York: Routledge.

Jungfraujoch (2017). Top of Europe. Promotional material. Retrieved from: https://www.jungfrau.ch/en-gb/jungfraujoch-top-of-europe/

Lorimer, H (2005). Cultural geography: The busyness of being 'more-than-representational.' *Progress in Human Geography*, 29(1), 83–94.

Macpherson, H (2010). Non-representational approaches to body-landscape relations. *Geography Compass*, 4(1), 1–13.

Magga, OH (2006). Diversity in Saami terminology for reindeer, snow, and ice. *International Social Science Journal*, 58(187), 25–34.

Nahrath, S and Stock, M (2012). Urbanité et tourisme: une relation à repenser. *Espaces et sociétés*, 151 (4), 7–14.

Rantala, O, Valtonne, A and Markuksela, V (2011). Materializing tourist weather: Ethnography on weather-wise wilderness guiding practices, *Journal of Material Culture*, 16(3), 285–300.

Schmidt, GM (2020). Vi er de sidste vinterbørn [We are the last winter children]. Retrieved from: https://politiken.dk/kultur/art7663704/Vi-er-de-sidste-vinterb%C3%B8rn Accessed 24 February 2020

Schneider, A (2009). Theme Park Europe: Transmission and transcultural images in popular Hindi cinema. In A Schneider and B Meissmann (eds.), *Transmission Image: Visual Translation and Cultural Agency*, (pp. 86–107).Newcastle Upon Tyne: Cambridge Scholars Publishing.

Spufford, F (2010). *I May Be Some Time*. London: Faber and Faber.

The Local (2019). Jungfrau visitor numbers reach new heights despite fewer skiers. Retrieved from: https://www.thelocal.ch/20190104/jungfrau-visitor-numbers-peak-in-2018-despite-falling-skiers

Thorpe, H and Rinehart, R (2010). Alternative sport and affect: Non-representational theory examined, *Sport in Society*, 13(7–8), 1268–1291.

Thrift, N (2008). *Non-representational Theory: Space, Politics, Affect*. New York: Routledge.

Vannini, P (n.d.). Non-representational theory and ethnographic research [Online]. Available from: http://ferrytales.innovativeethnographies.net/sites/default/files/Non-Representational%20Theory%20and%20Ethnographic%20Research.pdf. Accessed 1 November 2019.

Vannini, P (2015). Non-representational ethnography: New ways of animating lifeworlds. *Cultural Geographies*, 22(2), 317–327.

Vannini, P and Taggart, J (2014). Chapter 6: The day we drove on the ocean (and lived to tell the tale about it): Of deltas, ice roads, waterscapes and other meshwork. In J Anderson and K Peters (eds.), *Water Worlds: Human Geographies of the Ocean* (pp. 89–102). Farnham: Ashgate.

Vannini, P, Waskul, D and Gottschalk, S (2011). *The Senses in Self, Society, and Culture: A Sociology of the Senses.* New York: Routledge.

Xiao, H, Jafari, J, Cloke, P and Tribe, J (2013). Annals: 40–40 vision. *Annals of Tourism Research*, 40(1), 352–385.

9 Making the Santa Ana wind legible

The aeolian production of Los Angeles

Gareth Hoskins

Introduction

In January 1968 the journal *Professional Geographer* published Willis H. Miller's attempt to confirm a direct correlation between the Santa Ana wind of Southern California and criminal activity. Miller, an economic geographer at California Western University, compared homicide figures from the Los Angeles Police Department and County Sheriff's Office with United States Weather Bureau data in order to extend the resolution of the FBI's running presumption of a 'direct and significant relationship between seasons and the incidence of at least some types of crime' where summer produced more 'murder, forceable rape, robbery and aggravated assault' (Miller, 1968: 23). This chapter draws upon digital newspaper archives, and historical, meteorological and social science research, as well as literary sources in crime and romantic fiction to examine the Santa Ana wind as an object of enquiry and a force that cultivates negative emotional contagion. I provide a Santa Ana wind-centric history of Greater Los Angeles made up of two distinct modes of legibility, each with their own way of registering this wind's resistance to scrutiny and its agentive excess.

Alongside its more well-known and often tragic association with wildland fire, the Santa Ana wind has been productive in shaping the economic and emotional development of Greater Los Angeles. It has gained a reputation for altering mood and engendering social unrest in the areas through which it blows to the extent that like the French *Mistral*, the Swiss *Foehn*, the *Sirocco* of Italy, Canadian *Chinook* and *Sharav* of Israel, the Santa Ana wind is woven into the Los Angeles cultural fabric and civic infrastructure. The Santa Ana wind moves from the north east into the lowlands of Southern California around half a dozen times a year, mostly during the winter months when continental high pressure pushes dry air from the Mojave Desert towards low-pressure systems off the Pacific Coast. As it funnels through mountain passes and canyons and descends downslope, the air undergoes compressional warming and increases in speed – in rare cases up to 110 mph. The Santa Ana wind fits within the meteorological category of foehn-type or katabatic wind but is not sufficiently contained by these terms.

It is particularly vigorous through four channels: the San Gorgonio Pass, the Cajon Pass, the Newhall pass, and the Santa Ana Canyon (one explanation for its name), but its precise path is irregular and unpredictable. It is known as capricious, cruel, devilish, fickle, furious and ferocious. It howls through some valleys when others alongside are almost still. It is gusty and intermittent. It can change direction and swing around. Sometimes the Santa Ana wind remains benign in the upper atmosphere, uplifted by a ground-hugging onshore breeze, but then drops to cause havoc on an unsuspecting coast or on wildlands, foothills and ranges covered with housing.

Despite extensive investigation by the aerospace industry (Strange, 1936), atmospheric scientists, meteorologists and wildland fire researchers (Mensing et al., 1999; Raphael, 2003; Guzman-Morales et al., 2016; Rolinski et al., 2019), there is no agreement on the Santa Ana wind's precise mechanisms, nor on its regional extent. For some local meteorologists (for instance, Edinger, 1967), the Santa Ana wind is a fairly insignificant part of the weather regime. For others, it can describe any offshore wind that blows in California.

If the wind is illusive in theory it is even more difficult to predict and detect in practice. We do not know if the Santa Ana wind will blow, where it will blow or for how long. Predictive models and best-guess estimates based on its previous history generate such measures as the US Forest Service's Santa Ana Wildfire Threat Index (SAWTi), and over the last 50 years, as the threat to people and property from wind-blown fire has become more apparent, a defensive array of Remote Automated Weather Stations (RAWS) and high-definition video cameras with smoke detectors have been installed in remote wilderness areas throughout Southern California. But even with the live data streaming from high-sensitivity instruments, we cannot know the Santa Ana wind in its entirety since it is not the wind itself being measured but various detectable dimensions such as speed, direction, temperature, pressure, humidity, electrical charge and material composition. The Santa Ana wind, like all winds, is always more than a set of indices. It resists our attempts to make it intelligible and hold it to account.

This chapter works through two distinct efforts of legibility associated with the Santa Ana wind. On the one hand, there are the ever more elaborate scientific attempts to register, map and precisely define the Santa Ana wind as a meteorological and social-psychological object of concern. This extends to historical research and debate about the Santa Ana wind's etymology (Stephenson, 1943) and is characteristic of a modernist imperative to eradicate or instrumentalise uncertainty. The weather is frequently made calculable, for example, through the industries of risk, insurance and actuary. Pryke examines weather derivatives as 'geomoney' designed to 'capture and harness the vagaries of the so-called natural world and transform the associated risks into financial instruments ready for circulation' (2007: 578), while Szerszynski (2010: 22) discusses society's demand to tame the weather's 'material and semiotic unruliness' through 'instrumentation and standardized practices of measurement and statistical aggregation, [where the] weather [is] increasingly made to report as a "calculable coherence of forces'".

Willis H. Miller's attempt to evidence a relation between the Santa Ana wind and crime in downtown Los Angeles is among a small number of efforts to make the Santa Ana wind legible as a bio-meteorological or meteoropathological phenomenon. While the Santa Ana wind certainly has enough notoriety to interest anthropologists (Strauss, 2007) and receive mention in medical studies of other winds (Danon and Sulman, 1969), its behaviourial impacts are surprisingly under-scrutinised. In the field of finance economics though, Saporoschenko sought to link the Santa Ana wind and Southern California stock returns, but he concludes that 'no substantial evidence of the localized trading proxies capturing a weather-related stock effect is found' (2011: 693).

A second effort of legibility directly associated with the Santa Ana wind is driven not by modernist imperatives to epistemologically contain but by a need to open things up. Crime, romantic fiction and other forms of regional literary expression often use the wind's illusiveness as a source of enchantment and wonder; as a means of inviting subjection to elemental or planetary forces rather than promising mastery. In his analysis of anglophone literature set in Los Angeles, McClung (2002: 2) identifies this as a form of geographical orientation:

> Inconsistent testimonies about the climate are nearly as old as Los Angeles itself; they contribute to the uneasy overlay of conflicting mythologies on which the city and its chroniclers have fed. Such testimonies are efforts to make sense of the city in space and time by locating its present state, however incompletely seen or understood, as a point on a time line defined as much by myth as by history.

It is this mythological quality of the Santa Ana wind that enables writers to expand and intensify the Los Angeles imaginary.

These two efforts of legibility are both sense-making and place-making exercises and are political in a number of ways that too often go unremarked. First, they are political as a device of settler colonialism since they identify, name and imprint the wind with meanings and stories that supplant previous loric landscapes familiar to the more than 10,000 Indigenous Gabrielino-Tongva peoples almost eradicated by forced assimilation and genocide (Sawhney, 2002). Second, these efforts of legibility are political in their gendered and racialised encoding. The Santa Ana wind is often visually depicted as a white, bearded, God-like male blowing down on the earth below. In contrast, those who receive the wind (for instance, those who are most sensitive to its effects) are most commonly children, women, racialised groups and animals.

In what follows, I proceed through these modes of legibility in three sections that illustrate how we variously assign meaning to weather in a regional context. I start with a chronology of the Santa Ana wind as it appears in local newspapers to track its emergence as a disruptive feature of residential life in Los Angeles. Second, I consider its deployment in crime and romantic fiction where this wind's suite of affects is more precisely defined using sensory and erotic literary motifs. Third, I reflect on reports of the wind's agentic excess to set up a critique of Miller's wind-crime data which is presented within a broader context of early weather-determinism. I conclude the chapter with a cautionary note on the (mis)use of the Santa Ana wind as a scapegoat for structural inequality.

A chronology in newsprint

The wind is mentioned for the first time by name as 'the Santa Anas' on 15th November 1880, in the *Los Angeles Evening Express* in an article entitled 'The philosophy of sandstorms'. It is depicted again soon after in an 1883 history of San Bernardino County as 'disagreeable northerners... hot, parching winds from the desert, which, though seldom boisterous, are depressive and destructive, as they evaporate all moisture from vegetable and animal life, leaving everything scorched and parched up' (Lawton, 1965). From then on, it is referred to regularly within local newspapers as a fixture in the developing regional economy. Titles such as the *Riverside Daily Press*, *San Bernardino Sun*, *Los Angeles Herald*, *Corona Courier*, *Oxnard Daily Courier*, *Anaheim Gazette*, *Santa Ana Weekly Blade* and *Santa Ana Daily Register* noted how the Santa Ana wind destroyed walnut harvests and caused damage to the Valencia orange crop. It stripped bushes of autumn foliage; damaged commercial and residential property; tore down signage; flipped over trailers, campers and planes; and drove boats and barges ashore. It blew down oil derricks, disrupted operations at film studios and was mooted to be responsible for a number of peculiar events. In Anaheim in September 1894, for instance, it was blamed for unusual activity in real estate transfers. In 1912, its capacity to excite the hunger of fish was used to explain a record catch of longfin tuna off Longbeach, and in February 1918, it was blamed for killing mussels exposed at low tide on Newport Beach. The Santa Ana wind cleaned out a shoe store in Riverside, scattering shoes up and down the street; it was also blamed for prompting misbehaviour in classrooms, and for disrupting various hunting trips, flower shows and golf tournaments in towns throughout the Los Angeles Basin.

The Santa Ana wind often coincided with public events, lending them additional portent. In December 1892, its fury outside a courthouse drowned out an attorney's plea on behalf of an accused man. In 1899, it added to the sense of suffocation during a wait for the acquittal of a Mrs Cook. On 25th November, a national day of mourning after the assassination of President Kennedy, a Santa Ana wind hit the town of Redlands and loose fronds dropped from Mexican palms littering the streets. In November 1969, it set the scene for marchers protesting the Vietnam War, and in 1974, it accompanied feminist activist and attorney Flo Kennedy in a 'cyclone of outspoken' appeal for minority rights at the University of California campus in Riverside (Sun Telegram, January 26th, 1969: A-4). In these cases, the wind adds a layer of meaning either by providing commentary or in the enrolling of people into shared atmospheres.

A different notable characteristic of the Santa Ana wind for local newspapers is its sudden and unexpected appearance. It knocks fishermen off rocks, blows hikers off cliffs. In December 1937, it pushed Arnold Webster, a 'grip' for MGM studios, off the top of a 30-foot building while he was working on the sets of *Madelon*, and in February 1964, a sudden and 'howling' Santa Ana wind blew two Huntington Beach surfers five miles out to sea while people on the beach looked on powerless to intervene.

The Santa Ana wind's impact was not exclusively negative. In some cases, its appearance was reported as benign, even beneficial. Newspapers report it as removing fruit from trees to save the trouble of picking them and acknowledge its utility in drying grapes for the raisin industry. In an appendix to the Ninth Biennial Report of the State Board of Health of California for the Fiscal Years from June 30, 1884 to June 30, 1886. The Santa Ana winds are reported as 'health-giving' (Orme, 1886) and are celebrated frequently for their capacity to clear out smog. Specific benefits to health are noted in 1894 by a visitor to California reporting on his travels to readers of *The Red Cloud Chief*, a local news title in the Midwest:

> Every few weeks or so a hard wind called Santa Ana winds blow from the north east and causes the dirt to fly as thick as it does in Nebraska. The people of California dislike to have those storms. I believe they are of benefit to the country for they blow away the germs of different diseases brought by people who come there for their health. Also, the impurities caused by the decaying vegetation and leaves of fruit trees etc., which fall on damp ground caused by irrigating and causes the moss to grow on the buildings, fruit trees, fences and on the top of telegraph poles.
>
> (Foster, 1984: 1)

It is worth reflecting why newspapers constructed the Santa Ana wind as a singular object of concern. Its configuration as peculiar adds colour to local events and provides drama to otherwise mundane everyday activities, especially in the juxtaposition between acutely local and planetary scale weather. Repeated reference to the Santa Ana wind thus serves to cultivate a regional identity through a shared experience of the unique rhythms and patterns of a place so recently and rapidly inhabited by white settlers, and newspapers served as the means through which the public experience of the wind could be collectively apprehended.

The wind in literature

By the early 20th century, the Santa Ana wind's growing public notoriety was met with a legislative response from the Orange County and City of Santa Ana Chambers of Commerce who passed a resolution to rid their towns of its local association. An article entitled 'Would forget the "Santa Ana" winds, education campaign to cease use of name for "Northers"' relays the upset from directors at 'some other places that escape the odium and point the finger of scorn this way' (*San Bernardino Sun*, 1912). Indeed, the 'Santa Ana wind' label persisted and became further entrenched by its adoption within LA-specific genres of hard-boiled crime fiction emerging in the 1930s and 1940s.

The Santa Ana wind's association with malevolence and unease was popularised principally by Raymond Chandler and the short story *Red Wind* which was first published in the January 1938 issue of *Dime Detective Magazine*.

> There was a desert wind blowing that night. It was one of those hot dry Santa Anas that come down through the mountain passes and curl your hair and make your nerves jump and your skin itch. On nights like that every booze party ends in a fight. Meek little wives feel the edge of the carving knife and study their husband's neck. Anything can happen. You can even get a full glass of beer at a cocktail lounge.
>
> (1964: 69)

Red Wind sees detective John Dalmas witness a murder at a bar. Throughout the story of subsequent blackmail and corruption, the wind agitates. It opens and closes doors and booms 'like the sound of guns' (Chandler, 1964: 118). Later, when the murder is solved, 'the wind was all gone. It was soft, cool, a little foggy. The sky was close and comfortable and grey' (1964: 121). But even after the story's resolution, the Santa Ana wind leaves its mark to signal that the corruption and violence within the city was now a permanent feature:

> I went out of the bar without looking back at her, got into my car and drove west on Sunset and down all the way to the Coast Highway. Everywhere along the way gardens were full of withered and blackened leaves and flowers which the hot wind had burned.
>
> (1964: 124)

Following Chandler, the Santa Ana wind becomes a familiar motif if not quite a cliché in its repeated use within LA noir fiction. Dean Koontz (2006: 24) frequently employs it to generate suspense with an emphasis on animal affect:

> Eager breathing, hissing, and hungry panting arose at every vent in the eaves, as though the attic were a canary cage and the wind a voracious

cat. Such was the disquieting nature of a Santa Ana wind that even the spiders were agitated by it. They moved restlessly on their webs

And it is used similarly within the story 'Night Moves' by Tim Powers published in *Twilight Zone* magazine in 1988. The setting here is Guillermo's Todo Noche Cantina in Santa Margarita:

> [He] laid two quarters on the counter, got to his feet, and lumbered to the door. Outside, he tilted back his devastated hat and sniffed the night. It was the old desert wind all right, hinting of mesquite and sage, and he could feel the city shifting in its sleep - but tonight there was a taint on the wind, one that the old man smelled in his mind rather than in his nose, and he knew that something else had come into the city tonight too, something that stirred a different sort of thing than leaves and dust. This night felt flexed, stressed, like a sheet of glass being bent
>
> (Powers, 1988: 36)

Authors regularly note the Santa Ana wind's corporeal apprehension through olfactory aesthetics of aroma, smell and scent. Nigel Thrift's (2003: 10) discussion about the articulation of smell in contemporary society suggests why this might be so effective:

> aromas seem to escape our cognitive consciousness. They belong to a realm of 'peripheral' psychomotorial actions, an insistent substrate of incessant movement that makes up so much of what we are, but which we so often choose not to register as thought, even though the stamp of the impressions of this movement constantly influences us.

Clive Barker (2001: 1) develops this idea in the novel *Coldheart Canyon*:

> Whatever the truth of the matter, this much is certain: the Santa Anas are always baking hot, and often so heavily laden with perfume that it's as though they've picked up the scent of every blossom they've shaken on their way here. Every wild lilac and wild rose, every white sage and rank jimsonweed, every heliotrope and creosote bush: gathered them all up in their hot embrace and borne them into the hidden channel of Coldheart Canyon.

And for Koontz (1997: 146), again, the nose becomes an antenna registering the Santa Ana wind's capacity to mobilise the desert:

> It wasn't simply the baleful sound, like cries of an unearthly hunter and the unearthly beasts that it pursued, but also the subliminal alkaline

scent of the desert and a queer electrical charge different from those that other – less dry – winds imparted to the air.

In romantic fiction, the Santa Ana wind is employed to signal a different kind of tension – that of a repressed, or not so repressed, sexual desire. The wind's libidinal agency is exemplified in the opening passage of Elizabeth Lowell's 1979 novel *Golden Empire* summarised to readers as a story of 'three proud generations whose passion for the land was surpassed only by their passion for each other':

> The dry Santa Ana wind swept between the shoulders of the San Moreno Mountains and gusted down foothills cured gold by long days of California sun. The wind blew across the broad coastal plain with hot impatience until it plunged into narrow canyons leading from plains to sea. There in the canyons' shadowed places, the wind spent itself. When it finally emerged the wind was no more than the sigh breathed over the waters of the Blue Lagoon.
>
> (1979: 1)

Efforts to relate weather to sexual behaviour are fairly common. Leffingwell's 1892 link between illegitimacy and the influences of the seasons upon conduct, or recent clinical trials measuring frequency of sexual thoughts against temperature (Demir et al., 2016) are two examples from the positivist tradition, but weather-associated eroticism features in critical scholarship only very occasionally. Sturken (2001: 171), for example, identifies the repressed desire and displaced sexuality in contemporary television storm coverage which 'reveals itself to be a blend of science, eroticism, and sentiment, one that constructs citizens as participants in the production of weather narratives'. The sexualities untethered by the Santa Ana wind appear less coordinated or instrumental. However, its use in popular literature regularly invites a kind of volunteered subjection to planetary forces similar to that implied by Peter Weir's 1975 film *Picnic at Hanging Rock* (Aitken and Zonn, 1993) and adds another dimension to the more familiar gendering of nature as a passive recipient of male mastery. Hanging Rock, like the Santa Ana wind, becomes a repository for desire generated by a lack that can never be fulfilled.

Joan Didion implies something of this in her famous commentary on the Santa Ana wind in the essay 'Slouching Towards Bethlehem' (1968: 225): 'To live with the Santa Ana is to accept, consciously or unconsciously, a deeply mechanistic view of human behaviour'. There is here a form of submission to the Santa Ana wind; an embracing of its excess; a savouring of the thrill of our vulnerability that runs counter to rational scientific attempts to master the wind through instruments and calculation.

The Santa Ana wind is employed by writers to accomplish a number of metaphorical and narrative tasks. It emerged as a convenient way to set the scene and a literary shortcut to establish tension, and was developed latterly as a device to signal a sensorial, almost bestial, way of apprehending the world which occasionally, through submission or erotically charged capitulation, extends to hint at a more-than-organic collaborative sexual order (See Lingis, 2000).

Aeolian excess

The elemental politics of forces similar to those of the Santa Ana wind have been explored in recent work on cyclones (Last, 2015), 'geopower' (Grosz, 2017) and solar aesthetics (Engelmann and McCormack, 2018). My focus here is on an epistemological politics exposed by the wind's persistent withholding from technological scrutiny and representational containment. Resonant examples can be found in news coverage of the Santa Ana wind, as for instance, in this 1975 account of a Santa Ana wind-related misadventure at sea entitled 'Couple survives four-day ordeal on raft':

> Zovar said he and Miss Shepherd, a nurse, had been fishing about a quarter mile off the coast Sunday when strong winds blew their raft out to sea. "The Santa Ana wind came up in about one minute and we ripped loose from the kelp," Zovar said. "We yelled and screamed but nobody heard us. I had a pair of fins and I jumped overboard and kicked while my girlfriend rowed but it did nothing". Shortly after, they had their first of two encounters with sharks. "It was about half an hour after we tore loose" Zovar said. "We had two or three fish floating next to us on a stronger and the shark came up and grabbed them. We cut the stronger loose and just prayed". Zovar said the couple tried to get to shore but were forced back by the winds which pushed them 40 miles out to sea. Then the waves began to rise. On the morning of the third day the couple saw Anacapa Island about six miles away. "We tried to make it and we were blown back out again. Then yesterday we woke up in front of Santa Cruz Island and we tried again."
>
> (*San Bernardino Sun*, 1975, November 28)

This wind here is actively uncooperative. Veale, Endfield and Naylor (2014: 26) discuss the Helm Wind of Cross Fell in the Pennines, noting that 'winds have agency... they interfere with daily life'. The Santa Ana wind seems to possess a more specifically malevolent intentionality, which is also evident in the following example:

> Last Thursday evening one Louis D. Scarito was driving home on Etiwnda avenue, south-west of Kaiser Steel. He didn't make it. The fierce wind swept so much sand across the paved thoroughfare that his car got stuck. He was only a quarter of a mile from home. Then the sandblast held him captive inside while it ate the paint from his car and pitted the

glass. Rescued three hours later by the California Highway Patrol he said: "It was the worst experience of my life".

(Moore and Moore, 1957 *Redlands Daily Facts*, November 25)

The Santa Ana wind's growing profile as a regional problem through the 1960s aligned with the tail end of environmental determinism in geography where Willis H. Miller positioned local crime data alongside weather measurements obtained from a US Weather Bureau monitoring station located on the roof of the Department of Water and Power in downtown Los Angeles. His thesis was that

> If people in Southern California are uncomfortable and irritable on days when there is a pronounced Santa condition, it might be expected that an above normal number of crimes would be committed during those days. This study is a preliminary attempt to determine whether such a correlation exists in the Los Angeles Area.
>
> (1968: 23)

Miller confirmed that there had been 53 'Santa Ana wind days' during the study: 34 had an above-normal number of homicides, 3 had the normal number, and 16 had below the normal number. The total number of murders on 'Santa Ana wind days' was 58 compared to a normal 50.8 murders over an equivalent period without a Santa Ana wind. Miller (1968: 26) admitted that the results were 'believed to be indicative but not necessarily conclusive' and requested that the reader 'recognize that the time span for which data have been assembled is only two years, and there is no positive assurance that findings are not in part attributable to other currently unidentified factors'. With these qualifiers, Miller affirms the then-popular geographical and meteorological orthodoxy that a discrete psycho-physiological response to the weather can be positively identified.

One of the most confident and precise assertions of such a relationship appears in a study about the weather's impact on crime by Kropotkin in the 1870s: 'Take the average temperature of the month and multiply it by seven, then add the average humidity, multiply again by two and you will obtain the number of homicides that are committed during the month' (in Cohen, 1941: 31). By the early 20th century, the weather's influence on social behaviour was understood to be more complex. In particular, psychologist Edwin Dexter had been using statistical tests in his new laboratory at Columbia University to develop the field of behavioural bio-meteorology (Stewart, 2015). Dexter was part of new era of statistical positivism suspicious of broad links made between climate and human traits like national character, mental energy and economic progress that were common to Ellsworth Huntington, William Morris Davis, Friedrich Ratzel, Ellen Church Semple, and before them, Renaissance thinkers such as Kant, Montesquieu and Voltaire (see Harvey, 2000).

Dexter's 1904 book, *Weather Influences*, pinpoints the wind as a specific agent of violence: 'Whatever maybe our dislike for March hurricanes, the police judge does not profit from them. Our curves (Fig. 17) show that the mild winds of between 150 and 200 miles per day are the pugnacious ones' (1904: 152). He confidently explains this with a gas-based theory: 'This is easily accounted for by the fact that the usual excess of carbon monoxide in the air of cities is dispelled, and the increased oxygen brought in by winds heightens emotional and physical activity' (1904: 152).

In contrast to Dexter, Miller is not concerned with explanation. He is merely trying to evidence a relationship. And it is here in the story of Miller's data that we see the evasiveness of the Santa Ana wind as an epistemological object. The story of the data tells us a great deal about Los Angeles society at the time and the role of the geographer within it. It also provides a cautionary lesson about the risk of overreaching in our contemporary geographical analyses of the elemental that foreground the redistribution of agency and emphasise the role of the material in our more-than-human worlds.

If we put the vagaries of crime statistics to one side, Miller's research required two wind-related reductionisms to generate data that could be matched with LAPD statistics. The first involved deciding on a single point of measurement for the wind. Downtown Los Angeles was chosen not because it was especially windy [it is actually in the Santa Ana wind shadow] but because it was 'reasonably representative of the densely populated lowland areas of Los Angeles County' (Miller, 1968: 25). It is not too much of a mental leap to suggest that in a mid-1960s economic geography mindset, amidst a backdrop of heightened national concern about inner-city crime and urban rioting, the term 'densely populated' functions here as a euphemism for African-American. We are left to wonder what underlying ideas about race and behaviour Miller was attempting to imply by this study. We do know, however, that the Watts Riots, the largest incident of collective violence Los Angeles had then known, took place right in the middle of Miller's project and just a few blocks away from his monitoring station. The Santa Ana wind appeared not to be blowing at the time of the riots however, and so the events did not merit a mention in the article. Nevertheless, there was a great deal of professional energy invested in the idea that weather conditions, specifically 'thermal stress', were the trigger, even cause, of Africa-American violence (Maunder, 1970).

Once the location of measurement was confirmed, Miller's second reductionism involved deciding how to definitively count the Santa Ana wind. He determined that any day with a humidity value of 15% or less at noon would register as a 'Santa Ana wind day'. The direction and speed of the wind was recorded but did not seem to affect whether any particular day would qualify. Indeed, many days with the wind blowing in the opposite direction were counted as well as two days when there was no wind at all. The story of Miller's data reveals a great deal about how our epistemological struggles to isolate and enumerate the Santa Ana wind expose a naïve belief in the power of science to make things known, but they also expose a host of prevailing prejudices underpinning our research questions, project design and methodologies with regard to weather phenomena.

Conclusion

Making the Santa Ana wind legible remains a difficult endeavour that researchers continue to pursue with ever greater resolution (Raphael, 2003; Saporoschenko, 2011; Guzman-Morales et al., 2016; Rolinksi et al., 2019). But we still seem reluctant to relinquish an element of mystery evident in such glimpses as the local media's naming of Southern California Edison's leading meteorologist Tom Rolinski as the 'wildfire whisperer'. Efforts to make the Santa Ana wind legible have numerous political dimensions but given the limited space of this chapter, I have not been able to discuss them all. I have not, for example, included discussion on the socially uneven distribution of exposure to the Santa Ana wind where those with the least resources bear the greatest burden of risk (see Davis, 1995; Pyne, 2015). My focus, rather, has been on the politics of legibility and the epistemological challenges of understanding the social forces generated by and imposed upon meteorological systems more broadly in news reports, fiction and scientific research.

Two contrasting efforts of legibility show that we make sense of the Santa Ana wind either by containing it within a relational ontology (for instance, by seeking to definitively mark 'its' impact on other things) or by a creative-literary opening-up-to-the-wind as an inhuman nature or excessive planetary force. The danger with the former approach is that the Santa Ana wind can provide a simple answer to complex problems and divert blame away from a system that sustains fragile infrastructure and racially coded underinvestment, or away from a culture of structural misogyny by entertaining weather-related defence for domestic violence voiced as 'the Santa Ana wind made me do it' (Needham, 1988). Law enforcement agencies are regularly questioned and officers often oblige with quotes on the wind as a behavioural trigger, opining that 'when the Santa Ana wind blows, people get crazy', for example or 'we have always said when you get a Santa Ana wind watch out because you're going to have a lot of domestic arguments' (Lundah, 1983).

Research on the Santa Ana wind fits within a growing scholarly curiosity for the more-than-human, more-than-biological agencies, capacities and aesthetics of the atmospheric and elemental (Adey, 2013; Clark, 2017; Squire, 2016). As a weather mobility registered variously by European settlers looking to think themselves into the land, by atmospheric scientists looking to predict its appearance, and by writers looking to establish drama and tension, the Santa Ana wind has a productive ambiguity that warrants further study by geographers even as it remains perpetually illusive.

References

Adey, P (2013). Air/atmospheres of the megacity. *Theory, Culture & Society*, 30 (7–8), 291–308.

Aitken, SC and LE Zonn (1993). Weir (d) sex: representation of gender-environment relations in Peter Weir's Picnic at Hanging Rock and Gallipoli. *Environment and Planning D: Society and Space*, 11(2): 191–212.

Barker, C (2001). *Coldheart Canyon*. London: Harper Collins

Chandler, R (1938). *Trouble Is My Business* [1950 Edition], 69–124. London: Penguin.

Clark, N (2017). Politics of Strata. *Theory, Culture & Society*, 34(2–3): 211–231.

Cohen, J (1941). The geography of crime, *The Annals of the American Academy of Political and Social Science*, 217: 29–37.

Danon, A and FG Sulman (1969). Ionizing effect of winds of ill repute on serotonin metabolism. In *Biometeorology 4*, ed. SW Tromp and WH Weihe. *Supplement to International Journal of Biometeorology*, Vol. 13 Pt. II: 135–136.

Davis, M (1995). The case for letting Malibu burn. *Environmental History Review*, 19 (2): 1–36.

Demir, A, M Uslu and OE Arslan (2016). The effect of seasonal variation on sexual behaviors in males and its correlation with hormone levels: a prospective clinical trial. *Central European Journal of Urology*, 69(3): 285.

Dexter, EG (1904). *Weather Influences: An Empirical Study of the Mental and Physiological Effects of Definite Meteorological Conditions*. New York: The Macmillan Company

Didion, J (1968). *Slouching Towards Bethlehem*. New York: Farrar, Straus and Giroux.

Edinger, JG (1967). *Watching for the Wind: The Seen and Unseen Influences on Local Weather*. New York: Doubleday.

Engelmann, S and D McCormack (2018). Elemental aesthetics: on artistic experiments with solar energy. *Annals of the American Association of Geographers*, 108(1), 241–259.

Foster, WT (1894). Weather Forecasts. *The Red Cloud Chief Newspaper* published January 26, 1894. Red Cloud: Nebraska,

Grosz, E, K Yusoff and N Clark (2017). An interview with Elizabeth Grosz: geopower, inhumanism and the biopolitical. *Theory, Culture, Society*, 34(2–3):129–146

Guzman-Morales, J, A Gershunov, J Theiss, H Li and D Cayan (2016). Santa Ana winds of Southern California: their climatology, extremes, and behavior spanning six and a half decades. *Geophysics Research Letters*, 43, 2827–2834.

Harvey, D (2000). Cosmopolitanism and the banality of geographical evils. *Public Culture,* 12(2): 529–564.

Koontz, D (1997). *Sole Survivor*. London: Bantam Books.

Koontz, D (2006). *The Husband*. New York: Bantam Books.

Last, A (2015). Fruit of the cyclone: undoing geopolitics through geopoetics. *Geoforum,* 64: 56–64.

Lawton, HW. (1965) *Reproduction of Wallace W. Elliott's History of San Bernardino and San Diego Counties, California : with illustrations. 1883*. p. 100 Riverside: Riverside Museum Press, 1965.

Lingis, A (2000). *Dangerous Emotions*. Berkeley: University of California Press

Los Angeles Evening Express (1880). The philosophy of Sandstorms. November 15, 1880.

Lowell, E (1979). *Golden Empire.* New York: Fawcett Gold Medal Books

Lundah, M (1983). 'The Rampaging Santa Ana' *The San Bernardino County Sun,* October 2: 26.

Maunder, WJ (1970). *The Value of the Weather.* London: Methuen and Co. Ltd.

McClung, WA (2000). *Landscapes of desire: Anglo mythologies of Los Angeles.* Berkeley: University of California Press.

Mensing, SA, J Michaelsen and R Byrne (1999). A 560-year record of Santa Ana fires reconstructed from charcoal deposited in the Santa Barbara Basin, California. *Quaternary Research,* 51(3): 295–305.

Miller, WH (1968). Santa Ana winds and crime. *The Professional Geographer,* 20: 23–27

Moore, F and W Moore (1957). With a grain of salt. *Redlands Daily Facts,* November 25, p. 12

Needham, J (1988). The devil winds made me do it, Santa Anas are enough to make anyone's hair stand on end, *Los Angeles Times,* 12 March.

Orme, HS (1886). The climatology and diseases of Southern California. *Ninth Biennial Report of the State Board of Health of California for the Fiscal Years from June 30, 1884 to June 30,* 1886. Available from http://sandiegohealth.org/state/biennial/1884-1886_9th_Biennial_Report.pdf. (accessed July 8, 2019)

Powers, T (1988). *Night Moves.* Seattle: Axoloti Press.

Pryke, M (2007). Geomoney: an option on frost, going long on clouds. *Geoforum,* 38(3), 576–588.

Pyne, SJ (2015). *Between two fires: A fire history of contemporary America.* Tucson: University of Arizona Press.

Raphael, MN (2003). The Santa Ana winds of California. *Earth Interactions,* 7(8), 1–13.

Rolinski, T, Scott B Capps and Wei Zhuang (2019). Santa Ana winds: A descriptive climatology. *Weather and Forecasting* 34(2): 257–275.

San Bernardino Sun (1912). Would forget the "Santa Ana" winds. March 16, 1912.

San Bernardino Sun (1975). Couple survives 4-day ordeal on raft. November 28, 1975

Saporoschenko, A (2011). The effect of Santa Ana wind conditions and cloudiness on Southern California stock returns, *Applied Financial Economics,* 21(10), 683–694.

Sawhney, D (ed.). (2002). *Unmasking LA: Third Worlds and the City.* London: Springer.

Squire, R (2016). Rock, water, air and fire: Foregrounding the elements in the Gibraltar-Spain dispute. *Environment and Planning D: Society and Space,* 34(3): 545–563.

Stephenson, TE (1943). The Santa Ana wind. *California Folklore Quarterly,* 2(1): 35–40.

Strange, HE (1936). Study of the Santa Ana winds of the Los Angeles Basin. Masters Thesis. California Institute of Technology. Available from https://archive.org/details/studyofsantanwin109456578/page/n9/mode/2up. (accessed September 18, 2020).

Stewart, AE (2015). Edwin Grant Dexter: An early researcher in human behavioral biometeorology. *International Journal of Biometeorology,* 59(6): 745–758.

Strauss, S (2007). An ill wind: the Foehn in Leukerbad and beyond. *Journal of the Royal Anthropological Institute,* 13: 165–181.

Sturken, M (2001). Desiring the weather: El Nino, the media, and California identity. *Public Culture*, 13(2): 161–189.

Szerszynski, B (2010). Reading and writing the weather. *Theory, Culture & Society*, 27(2–3): 9–30.

Thrift, N (2003). All nose. *The Handbook of Cultural Geography*, in Anderson, K, Domosh, M, Pile, S, and Thrift, N (eds.), *Handbook of cultural geography*. London: Sage, pp. 9–14.

Veale, L, G Endfield and S Naylor (2014). Knowing weather in place: the Helm wind of cross fell. *Journal of Historical Geography*, 45: 25–37.

10 Seeing with Australian light

Representations and landscapes

Tim Edensor

Introduction

Our experience of the world is thoroughly shaped by luminosity, shadow and darkness, agencies that circumscribe what we can see, how we feel and sense, and how we make meaning with the light and the materiality with which we sense. As Mark Tredinnick (2013: 13) asserts, weather 'is the oldest story in the world – one we want to keep on telling each other when we meet, as though it were part of who we are'. Indeed, observations on the qualities of light seep into everyday conversational exchanges: 'It's bright today, isn't it'; 'Oh, it's awfully gloomy'; 'Where's the sun gone?' Thus light, as an integral element of the weather, shapes the qualities of place, our sense of belonging and our lay geographical knowledge.

Wherever humans live, we inhabit realms in which variegated patterns of sunlight, shade and darkness tone landscape with colours, shadows and textures that ceaselessly reconfigure our apprehension and lure our attention. We are, as Tim Ingold (2008) emphasises, always immersed in the currents of a world in formation, and this dynamic play of light is an integral part of the everyday affective and sensory experience through which we become attuned to place. Sedimented in mundane sensory experience, the luminous qualities of place are unreflexively apprehended most of the time, but they occasionally surge to attention when we find ourselves in an unfamiliar location or when they change. As I will demonstrate, arriving in an unfamiliar Australian setting attuned me to the very different manifestations of sunlight that shone across space, triggering sensory experiences that greatly diverged from my habitual encounters with British light. Disruptive transformations in the usual weather are exemplified by the highly colourful sunsets across northern Europe in 2010 that were caused by the eruption of the Icelandic volcano Eyjafjallajökull and the recent swathe of bush fires that have impacted upon the ways in which Australians see with light and breathe air. Yet though these incidents in which weather dramatically deviates for usual experience are becoming more common with environmental and climate change, weather still often remains tethered to notions of spatial belonging.

The spatial scales through which we identify distinctive forms of weather vary from the local and the regional to the national. While weather is experienced in local contexts, and may possess very distinctively local qualities, large weather systems also predominate over larger areas, perhaps shaped by sea or mountains. Whatever the case, although many weather systems are not coterminous with national or regional borders, national or regional qualities are nonetheless assigned through cultural discourses and representations that make sense of shared lived experience of weather. As Henry Plummer (2012: 6–7). notes, at a supranational scale, a distinctive Nordic light radiates across Scandinavian landscapes, despite their diverse topographies and ecologies, wherein 'their skies share a subdued light that imbues the entire region with mystery'. Winters are characterised by the 'low slant of the sun... long shadows and strikingly refracted colours' that pervade the winter months of these landscapes, while on midsummer evenings 'the sun dissolves into an unreal haze that bathes the land in a fairy-like glow, its colours strangely muted and blurred'. Plummer considers how such effects have 'permeated the arts' in these countries, but they also resonate with lived experience, for such patterns of light and dark shape cultural rhythms across the Nordic region, influencing when social practices take place across the year.

National identities are also powerfully informed by notions of 'national' weather that permeate common apprehension, Indeed, as Melissa Miles (2013: 261) reveals, 20 nations adorn their flags with images of the sun, underpinning 'the diverse, deep, and widespread connections forged between light, the sun, and national identities'. In this chapter, I focus on the qualities of Australian light and the ways in which a sense of this luminosity is acquired in everyday experience and is consolidated by influential cultural and creative representations. First though, I discuss how light is perceived, conceived and experienced by humans.

Seeing with light and landscape

We see with light according to four factors: first, the specific visual apparatus of the human visual system; second, the specific qualities of the light that shines on place; third, the particular materialities in the landscape upon which light shines; and fourth, the cultural conventions through which we discern light and make sense of it.

Light constitutes a form of radiant energy that is not perceived in itself, but, rather, through the visual perception of the diverse colours and intensities it produces as it shines on space. Light enters the human eye and enables us to visually discern selective parts of the electromagnetic spectrum, as the eye's convex lens focuses light to produce an inverted image of a scene on the retina. The iris expands and contracts in controlling the amount of light admitted. The image is sent from the optic nerve via electrical impulses to the brain, which processes this information, controlling selection of the infinite elements in the field of view.

The human capacities of vision through which we see with light are distinctive. Our night vision is quite efficacious in comparison to other creatures, but we require a period of around 20 minutes for the rod cells in our eyes to adapt to the lower levels of light, allowing us to discern gloomy landscapes with greater acuity (Edensor, 2013). We see colour well but cannot see particularly far with any accuracy, as can birds that can discern prey from a great distance because of the higher quantity of light receptors in their retina. And unlike many birds of prey, we are unable to see with ultraviolet light. Yet in any case, we find it difficult to look directly at things in a concentrated fashion but rather our eye roams and ranges across landscapes, and we typically move along a path of observation that shifts the perspective we adopt (Ingold, 2016).

The entanglement with our perceptual apparatus underpins how light circulates between that which we commonly assign as external and internal. And this discloses that notions that construe a separate viewer gazing upon an objective landscape are misplaced, for the light that reflects, is absorbed and deflected across space, also flows through us. Immersed in realms of light, weather and earthliness, any sense that we are embodied human entities separated from the landscape is deceptive. This is partly forged by persistent conventions of cultural representation and the persistent norms about how to look upon landscape, assess, understand and characterise it (Edensor, 2017a). Yet, landscape is neither object nor subject but blends distinctions between the looked upon and the onlooker, and both humans and landscapes continuously emerge, partly through their unceasing interaction with each other (Wylie, 2006).

For instance, as a phenomenon of light, colour bestows a particular quality to the sensible world; as Ingold (2010: s129) considers, to see the sky 'is to see *in* its light; therefore, since the sky is blue we see in its blueness'. As he further emphasises, colour is not a mere 'adornment, conferring an outer garb to thought, but the very milieu in which thought occurs' (Ingold, 2016: 215). Light is thus the medium with which we see:

> Seeing with sunbeams is like feeling the wind: it is an affective mingling of our own awareness with the turbulence and pulsations of the medium in which we are immersed. No more than the wind is the sun given to us as an object of perception. It rather gets inside and saturates our consciousness to the extent that it is constitutive of our own capacity to see or feel.
>
> (Ingold, 2016: 222)

This does not minimise the ways in which we interpret landscape, for instance, with regard to the cultural significations accorded to particular colours. For instance, the rising sun can produce excited anticipation of the day ahead, the setting sun frequently generates a melancholic sense of day's dawning, while a full moon may solicit romantic thoughts. Such cultural associations become embedded in affective and emotional life, and are deployed in religious conceptions and architectural designs.

In further developing an understanding of how we see with light and landscape, Alphonso Lingis (1998) identifies how changing *levels* of light, characterised by depth of field and brightness, continuously play across space, forming a fluid realm within which we see things and with which we continuously adjust. Shifting tones, shades and intensities lure our gaze towards particular colours and luminous areas and engender continuous (re)attunement. Yet, nonhumans also react to these changing levels of light: plants turn their leaves to the sun, bats shrink from daylight, cows seek shade and lizards bask on sunlit rocks. It is thus salient to consider how light charges the landscape with a vitality that is signified by the numerous nonhuman organisms that occupy it.

A sensory experience of particular landscapes is also shaped by the patterns of light, darkness, radiance and intensity that play across space, along with the changing angles of the sun's rays, seasonality and diurnal passage. Positions of latitude and longitude are critical here, but they are ameliorated by large weather systems and flows that impact upon the qualities of light. In this context, Australia receives much more solar radiation than California or the Mediterranean, a luminosity that is akin in intensity only to Tibet and central Africa. Besides visible light, this includes infrared radiation that generates heat and a high proportion of ultraviolet radiation that is responsible for sunburn and the nation's high incidence of skin cancer, damage to human health that is exacerbated by the hole in the ozone layer. In addition, being in the southern hemisphere, shadows extend in the opposite direction over the course of the day to those in the northern hemisphere. Location also determines whether and when levels of light are sustained, producing rather unvarying spells of greyness or bright sun, or whether these are more momentary, as when clouds rapidly scud across the sky and tone landscapes with swift oscillations between brightness and shade.

Our perceptual experience of light is also highly conditioned by the affordances of place, the ways in which light interacts with the distinctive surfaces, materialities and objects upon which it reflects, deflects and is absorbed, and the capacities of landforms, fixtures, buildings and living things to attract and block luminosity. For example, light is reflected by water, absorbed by the pigment cells of plants and obstructed by large trees. Moreover, the distinctive ways in which things and material masses block the varying luminosity of the sun's rays create distinctive forms of shadow and shade. The experience of light is also shaped by how it inflects the world with colour. As Diane Young submits, colours 'animate things in a variety of ways, evoking space, emitting brilliance, endowing things with an aura of energy or light' (2006: 173) across time and space, with the sunlight bestowing the material surfaces of place with distinctive hues that change according to season, time of day and the light's intensity. These effects of light can be harnessed in architectural design. Architect Juhaani Pallasmaa

(2012) focuses on how materials produce distinctive effects when deployed in buildings, creating moods and aesthetic effects according to how they interact with light. This is exemplified by the ways in which the white marble of the Taj Mahal changes colour and tone throughout the day in accordance with the shifting qualities of the light, radiating brilliantly in the midday sun, becoming rose-tinted at dawn and taking on a milky translucent glow at twilight. In considering how light colours cities due to their specific material constitution, the grey granite tones of Aberdeen, Scotland; the mellow yellow colours of the sandstone of Bath, England; and the russet hues of Siena, Italy, evidence a highly distinctive appearance.

In combining the qualities of light with material affordances, a walk across a moor in the Peak District National Park revealed a variant of a distinctively British encounter with light and landscape (Edensor, 2017b), which according to Greenlaw (2006), is typified by cloudiness, mild shadows and weak sunlight that produce distinctive tonal atmospheres typified by subtle and ever-changing patterns of light and shadow. An aesthetically compelling account of the diverse effects of particular British landscapes under the continuously changing and seasonably variable of twilight is also provided by Peter Davidson (2015).

Finally, it is also crucial, as David Howes (2010) insists, that despite the immersive immediacy of sense experience, this should not distract us from the ways in which this is culturally constructed. For distinctive cultures of looking ensure that seeing with light does not provide unmediated access to reality; rather 'seeing involves movement, intention, memory, and imagination' (Macpherson, 2009: 1049). For instance, in considering the play of light in colouring space, purely scientific conceptions that construe the experience of colour as a mechanical or psychological response and colours as purely objective qualities cannot get at 'the emotion and desire, the sensuality and danger, and hence the expressive potential that colours possess' (Young, 2006: 174). Accordingly, our affective engagement with light – as with all weather effects – is informed by the particular cultural conventions that inform how we look at and understand what we see: the play of light provokes onlookers into making sense of landscape, as I depict below according to distinctively Australian perspectives.

For example, we are drawn to views to which we have been culturally attuned, as with the enaction of viewing landscape in accordance with the conventions of romantic gazing (Urry and Larsen, 2011) from the peak of a mountain, through which we may interpret a patch of sunlight as 'sublime' or a collection of trees, cottages and streams as 'picturesque'. Strong colours too, elicit divergent cultural responses, as Diana Young (2011: 367) exemplifies in a nuanced account of the multifarious meanings of colour amongst the Aboriginal Anangu culture, discussing how bright hues are regarded as sensuously and affectively enervating as well as symbolising vitality and 'a manifestation of Ancestral energy and presence'. On the other hand, David Batchelor (2000) points to the Western suspicion of gaudy colours as part of a contemporary cultural preference for muted colours and widespread *chromophobia*, wherein landscapes are managed to ensure that light cannot trigger brightness in ways conceived as vulgar and intrusive.

I now turn to consider how we see with Australian light, with its particular qualities, the distinctive materialities upon which it shines and the cultural representations that have emerged. I firstly discuss how particular spaces of light have been artistically represented before endeavouring to identify certain particular attributes of the summer light of Australian landscapes, specifically in the city of Melbourne and in its rural surroundings.

Representing the light of the Australian landscape

I emphasise that I am not making any general claim about the qualities of Australian light as a whole, which is varied and multiple according to place and region, and throughout the day, season and year. Nonetheless, I submit that there are consistencies of recurrent experiences of light, together with the cultural nuances rendered through representations that make discussion of Australian light pertinent. As Melissa Miles (2013: 228) states, 'Australians came to invest heavily in the uniqueness of their sunlight as a symbol of collective unity and identity'. I now focus on certain renowned cultural representations in painting and poetry that sustain the notion that light is a marker of national identity before turning to more critical, contemporary representations in painting and photography that critically interrogate these widespread representational tropes but nevertheless also reinforce a sense of Australian light's distinctiveness. Of course, these settler colonial assertions diverge from Aboriginal understandings and representations of light, as inferred above, as well as those of other creative practitioners, yet they testify to the continuing common-sense force that Australian light is distinctive.

In considering the ways in which the light of Australian landscapes has been represented in painting, I focus on those produced by the Heidelberg School, that emerged in the eponymous north Melbourne suburb in which they briefly lived and painted in the late 19th century. Primarily, they created plein air paintings of Victorian, rural settings and coasts around Melbourne, and in Gippsland and Yarra regions. The naturalist, impressionistic landscapes by Tom Roberts, Arthur Streeton, Charles Conder and Frederick McCubbin profoundly articulate a relationship to Australian luminosity, with brilliant skies, luminescent bodies of water, bleached fields, and vibrant yellows, blues and whites that captured the powerful light of Victorian summers. These depictions were aligned with an emerging sense of national identity that sought to distinguish itself from the British artistic sensibilities that predominated under earlier phases of settler colonialism. Most paintings of Australian landscapes created by settler colonialists before the 1880s were suffused with muted colours, cloud-filled skies and light that played gently across space, scenes more redolent of European landscapes. Perhaps in encountering the unfamiliar bush landscapes and the harsh light in which they were often bathed, these artists sought to make sense of them by representing them in familiar, coded ways. Although they certainly capture the luminosity of the light of these landscapes, the Heidelberg School's paintings have been appositely critiqued since they participate in propagating the persistent nation-making myth that rural Australia is primarily terra nullius, a land that is hard won from the elements by toil, patience and stoicism, as reflected in the iconic figures of the bushman and the digger that feature in these paintings. There are no Aboriginal people, except as 'exotic' elements of local colour, and neither are there any significant signs of their stewardship of the land.

Complementing the nationalist impetus to apprehend and appreciate Australian landscapes as vibrantly luminous to distinguish them from the cultivated, managed scenes of muted greenery that pervaded representations of rural England, Dorothea McKellar's renowned poem from 1908, *My Country*, includes the following verse,

> The love of field and coppice
> Of green and shaded lanes,
> Of ordered woods and gardens
> Is running in your veins.
> Strong love of grey-blue distance,
> Brown streams and soft, dim skies
> I know, but cannot share it,
> My love is otherwise.

In later verses, McKellar goes on to detail landscape features of 'sweeping plains', 'ragged mountain ranges', 'droughts and flooding rains', 'stark white ring-barked forests', 'hot gold flush of noon' and 'pitiless blue sky', features that distinguish an Australian luminosity from the landscapes of England.

These notions of a merciless light are especially vivid in Barbara Bolt's account of her painting practices in Kalgoorie, Western Australia, a marginal desert landscape so saturated with light that vision is impaired and, consequently, so are attempts to represent its qualities. In conceiving the effects of this dazzling radiance, Bolt (2000: 203) questions the reiterative enlightenment conception, that in the 'heliophilia of Enlightenment thinking, the relationship between light, knowledge, and truth is assumed, and it is through vision that this nexus is achieved'. Thus, the light that falls on the world reveals and renders legible the reality of that which we look upon, banishing the ignorance caused by metaphorical and actual darkness. Such understandings, Bolt suggests, are predicated upon a European light that discloses details to the onlooker, reveals what is there to be seen. By contrast, in Australia, in the glare of the midday sun, a shadowless realm, nothing is revealed: 'too much light on the matter sheds no light on the matter' (205). In seeking to paint the Kalgoorie, the intense glare and irregularity created a landscape that was 'so fractured and messy that no form emerged' (206), no clear distinction between foreground and background. Light cannot here inform any clear representational strategy to 'capture' the landscape, and for Bolt, this glare solicited a 'downward look and an attention to the patterns and rhythms of the ground' in painting the landscape (209).

Besides painting, Rod Giblett (2007) claims that landscape photography has played a key role in generating notions of Australian-ness, notably the wilderness photographs that depict huge landscapes bathed in light and harsh shadow, and Max Dupain's renowned image from 1935, 'Sunbaker'. Indeed, from the late 19th century, the sun has loomed large as a powerful symbol of national identity, signifying the vitality of the nation, distinguishing it from the grey British motherland and representationally and discursively domesticating an often-forbidding land. Melissa Miles (2013: 263) exemplifies these ploys to mark out Australia's light as distinctive in the Sydney Camera Circle's pledge to 'show our own Australia in terms of sunlight rather than those of greyness and dismal shadows'– the endeavour to distinguish this from UK is clear. Yet, echoing Barbara Bolt, she points to a desire in contemporary Australian photography 'to court the dazzling, disruptive, and volatile aspects of light', 'a suspicion of light's ability to serve as an agent of truth and revelation in photographic representation' (Miles, 2008: 221). In characterising this approach to representing light, she discusses Danielle Thompson's *Halo #1* in which the light that filters through the trees assumes the form of sharp luminous splinters that cut through and across the landscape. In providing a sense of how such powerful light is experienced, Miles identifies 'how we idealise and romanticise a light that can also obscure, burn, bleed, pierce, and blind in photographic practice' (234). She also discusses Paul Fusinato's Sun Series in which the 'sun refuses to render its form as a 'stable, legible, and knowable entity' (226). In making serial images of the sun by directly pointing his camera towards it, Fusinato both highlights its changing dynamism in the 'interaction between the photographer, his camera, the weather, and the sun', and underlines an inability to represent the varying shades of intensity (224) that shape many of Australians' encounters with light. Both these photographers conjure up how dazzling light immanently tinctures experience, cajoling and repelling the gaze moment by moment. Such fragmentary experiences shift moods and feelings, and call for continuous adaptation to the changing levels of light.

Experiencing light in the Victorian summer

Having discussed selective creative representations that offer interpretations of the distinctiveness of Australian light, I now seek to identify the effects and qualities of summer light in two different landscapes, south Victorian rural landscapes and the urban landscape of Melbourne. Arriving in Melbourne in summer three years ago, as somebody habitually attuned to the pale, diffuse light, grey skies and vapid shadows of northern England referred to above, I immediately became aware of the very different effects of light on the Australian landscape. Moreover, in common with other residents of Victoria, I developed habits to manage interaction with the sunlight and heat. Before leaving an interior environment to walk outside, most citizens equip themselves with sunglasses to minimise the glare and deploy hats to shield their heads against the sun. Indeed, exposure to the dazzling sunlight has potentially harmful impacts, notably the high rates of skin cancer in Australia, as well as less severe discomforts of impeded vision, excessive sweating, headaches and thirst. The intensity of the light also determines whether people decide to walk or cycle: too much heat or brightness deter such mobile ventures (see Simpson, 2019).

In rural Victoria, when the sun is at its height, light suffuses landscapes, making colours especially luminous in both foreground and background. From a distance, expanses of sand radiate with an intense white as wet patches of beach reflect vibrant skies, azure water sparkles in seas and lakes, cultivated fields and foliage are clad in luminous tones of green, and skies glow with a concentrated blue. Stretching into the distance are the ubiquitous expanses of pale-yellow grasslands. Close at hand, parched ground is covered in earthy reds and browns; dry leaves and grass bestow a bronzed hue across surfaces that are interspersed with shoots of gleaming green grass; almost fluorescent rock pools are garishly adorned with splashes of blues, greens and reds; and flowers radiate brilliantly against darker greens. Roadside advertisements, painted houses and garden ornaments appear lurid in the intense glare. Tree-filled landscapes scorched by bush fires are replete with black, burned trees that stand out prominently against the bright expanses in which they are rooted, while trees that have been bleached white by the sun also reverberate against fiercely lit backdrops. Vegetation is attuned to the weather, with the blazing light reflected on the floors of eucalyptus forests where stripy patterns of shade contrast with the blotchier, sparser shadows of broad leaf woodland, while glossy leaves glisten in the sun.

When the angle of the sun lowers towards the evening, the landscape becomes endowed with a phantasmagorical appearance, as the light shines fervently off all those planes and surfaces that face it, bringing out the minutest features of texture, colour and form, revealing details that usually remain undiscerned by vision. The vibrant light also produces vividly clear silhouettes in contrast to a blue or orange sky with telegraph poles, rock formations and trees marked vividly against horizons. In forests, looking up, the complex geometrical forms of tree ferns are etched against the sky, and in pastureland, trees offer deep patches of shade in which the silhouetted bodies of cows rest. As the end of the day draws near, effulgent sunsets of radiant intensity bathe the land in limpid tones of gold and blue, with firmaments replete with wispy white clouds or glowing, rippling mackerel skies.

This luminosity shines through atmospheres that are bereft of thick particles that reduce light intensity and clarity, in contradistinction to the skies of northern Europe. to which I am habituated, that are usually saturated with moisture and various particulate matter. Yet, at certain times of the year, a combination of pollen allergen particles released into the air from North Victorian grasslands, sweeping winds, hot temperatures, high air moisture and humidity can create severe weather conditions for those suffering from asthma, hay fever and other respiratory ailments. On 21st November 2016, as the temperature reached 35 degrees Celsius, several deaths were caused, a particularly severe result of what is a regular annual occurrence (ABC News, 2016). More recently, the air quality, and hence the quality of the sunlight that shone across large areas of South Victoria was dramatically affected by the smoke that travelled from bush fires in the south east of the state and from larger blazes in New South Wales. This intense smoke pollution had deleterious effects on health and, at the time of writing, persists in milder conditions, greatly limiting visibility and shrouding the landscape in a haze that diverges from the usual summer radiance.

Seeing with the light and shadows of Melbourne

Moving away from the rural realm, the vibrant summer light produces distinctive expressions of light and shadow that play across the dappled urban landscape of Melbourne. The blue Dandenong Ranges ring the east of the city while looking towards the sharply delineated forms of the ever-loftier city centre, high rise towers and cranes create prominent silhouettes against the radiant sky.

More broadly, the light imbues Melbourne with a polychromatic colour palette that possesses certain recurrent tones. The ubiquitous local basalt, or 'bluestone', absorbs sunlight to provide deep, dark hues that stand out against lighter colours and vibrantly reflect the sky when it rains (Edensor, 2020). By contrast, the mellow tones of the local Heatherlie sandstone that clads several of Melbourne's most prestigious buildings reflects the sun in a glowing, soft yellow-brown. The city centre influx of towers built with concrete, steel and glass, and highly polished granite radiate and gleam, while recent intrusions of polychromatic glass and copper reflect light in shifting multi-hued ways. The 47 massive, slab-like concrete, high-rise towers constructed by the Housing Commission of Victoria in the 1960s assert their presence in grey and brown, with occasional splashes of red or cream.

The intense sunlight creates intense contrasts, sharp lines, and clear-cut areas of shade. Blinding glints of sunlight reflect off the metallic bodies of cars and trams. The luminosity also solicits an enhanced grasp of the form of waste bins, streetlights, railings, chairs, tables and street signs that stand out against stone, concrete and asphalt backdrops. These everyday fixtures are also enhanced by the highly delineated shadows that stretch behind them. When the angle of the sun becomes acute, shadows become distorted or fantastically lengthened, and people, too, produce elongated shadows as they walk upon on flat, wide roads and across squares. Indeed, in these brilliantly illumined realms we continuously encounter our own shadows. The angle of the light can conjure up fantastical shadowy selves but shadows also confirm that we are an integral part of the landscape, one of the infinite elements that interacts with the circulating vitality of light, air and matter. Besides discrete bodies and fixtures, the shadows cast across vertical and horizontal surfaces bring surrounding elements of the environment into play – blotches of shade cast by trees, the inadvertent designs of electric and telegraph wires, the gridded patterns made by the mesh of fences and ironwork of gates. Most distinctive are the shadows produced by the light that beams upon Melbourne's 19th-century wrought ironwork, amplifying the intricacies of this unique feature (Edensor and Hughes, 2019).

Melbourne's central area and inner suburbs are amply provided with green spaces designed to create plentiful areas of shade in which inhabitants might escape the glare of the sun. Following 19th-century fashions in British park design, the large parks were well stocked with deciduous European oaks, planes and elms, but also the dense indigenous evergreen Moreton Bay figs, large trees that continue to provide deep shade that contrast with the aforementioned streaky shadowy patterns of native eucalypt woodland. In the summer heat, Melbournians seek respite underneath these trees, reclining and sitting on the ground. These patterns of shade also shape the paths that pedestrians choose to take, where canopies appended to buildings or the thick foliage of trees on city-centre sidewalks provide shade that lure crowds not evident on sides of the road that lack such shady protection. In parks, large numbers of picnickers and gathering of people eating lunch congregate under the thick shadow of plane and elm trees.

Architects and planners also respond to the dazzling light by creating structures that offer shade and that design with light to produce distinctive patterns of light that extend below installations and across facades. The 19th-century canopies that line rows of city-centre shops, many of which have been disassembled, have recently been complemented by a range of pavilions and canopies that create dramatic, sometimes elaborate patterns of light and shadow across floor surfaces. For instance, each summer, a new pavilion is temporarily installed in the Queen Victoria Gardens on Melbourne's South Bank where it serves as a venue for concerts, workshops and lectures. Each unique pavilion is subsequently re-erected in a different part of the city at the end of summer where it continues to blur distinctions between outside and inside and work with the play of light and shadow. In addition, contemporary high-rise office and residential blocks are being designed so that when the sun is bright, features such as sills, decorative brickwork and window shades produce shifting, geometric shadowy patterns across their facades. These recent examples follow from the ways in which architects through the ages have sought to inventively produce different aesthetic effects through manipulating light and shadow (Kite, 2017). Such manipulations not only apply to buildings, for as Böhme (1993) discusses, 18th- and 19th-century arboreal landscapes were orchestrated by landscape gardeners who 'tuned' space by managing the levels of light that filtered through woodland canopies on country estates.

Both the rural and urban light and shadow I discuss here are part of the ways in which we come to know weather, through the compilation of successive events, sensing the regularities of place through time and becoming attuned and oriented to these effects, and to which our seeing bodies become habituated. We come to know the places and moments in which glare, shade and heat shape apprehension of the environments through which we move and which guide our path-making. In this sense, we come to know light through movement, as Tim Ingold (2010) insists, moving across the shifting seasonal dispersal of luminosity and in accordance with the ways in which light inflects the specific materialities of place. This is exemplified in a particular event that has recently been staged twice annually in Melbourne, 'Melbhenge', where the light that radiates across the city creates a spectacle and a sense of occasion.

Early in one evening in February and October, several hundred people gather on the steps of the Victorian Parliament Building on Spring Street, many equipped with cameras and tripods. The view from the steps offers the throng a direct view of the undeviating linear mile-long stretch of Bourke Street, part of the Hoddle Grid, the rectilinear mesh of streets that form Melbourne's CBD designed by town planner Robert Hoddle in 1837. Named after the solstice celebration at the renowned ancient British site of Stonehenge, Melbhenge occurs when the setting sun descends to the west directly between the large towers that line many of the streets on the Hoddle Grid. Confined by the concrete, canyon-like structure of the street's monolithic tower blocks, the sun casts a radiant glow along the length of Bourke Street, brilliantly illuminating the asphalt surface and radiating along the gleaming steel tram tracks that line it.

Conclusion

In this chapter, I have emphasised that we can only perceive with light according to our distinctively human visual capabilities, capacities that diverge with those of nonhumans. We see with the particular qualities of the light that shines upon place, which depends upon a variety of factors including longitude and latitude, seasonality and air quality, elements that shape how the landscapes of South Victoria and Melbourne may be apprehended. Also critical are the materialities that deflect, absorb and reflect light in distinctive ways, contributing to the colour, brightness and shadowy qualities of place, as I have also demonstrated with the distinctive flora, architectures and objects that typically characterise rural and urban landscapes of Victoria. I contend that these effects are indeed distinctive and play across the materialities of landscape in distinctive ways. Yet, I have also stressed that understandings of weather – in this case, of light – are grounded in shared habits and conversations that domesticate everyday experience and make it knowable. Weather is thus invariably culturally interpreted, and these place-specific, situated sensory attunements, affects and meanings also resonate in artistic representations. Here, I have discussed how Australian painters, poets and photographers have sought to portray the luminous specificities of landscapes as part of a developing national identity, but also how these representations have been challenged by artistic works that focus on the dazzling, incapacitating qualities that render such spaces difficult to depict. As someone unhabituated to these effects, and having been immersed in differently toned landscapes for most of my life, Australian light and the landscapes it radiates upon appeared to me to offer a dramatic contrast to my usual encounters with space and sunlight. For me, this reveals how although for most inhabitants, sunlight may usually be unreflexively experienced, the qualities of tone, intensity, reflection, colour and shadow that it bestows upon place are integral to everyday experience.

References

ABC News (2016) '"Thunderstorm asthma": two die after Melbourne storm causes spike in respiratory problems', 22 November. https://www.abc.net.au/news/2016-11-22/two-die-in-thunderstorm-asthma-emergency-in-melbourne/8044558 (accessed 7 March 2020).

Batchelor, D (2000) *Chromophobia*. London: Reaktion.

Böhme, G (1993) 'Atmosphere as the fundamental concept of a new aesthetics', *Thesis Eleven*, 36(1): 113–126.

Bolt, B (2000) 'Shedding light for the matter', *Hypatia*, 15(2): 202–216.

Davidson, P (2015) *The Last of the Light: About Twilight*. London: Reaktion Books.

Edensor, T (2013) 'Reconnecting with darkness: experiencing landscapes and sites of gloom', *Social and Cultural Geography*, 14(4): 446–465.

Edensor, T (2017a) *From Light to Dark: Daylight, Illumination and Gloom*. Minneapolis: Minnesota University Press.

Edensor, T (2017b) 'Seeing with light and landscape: a walk around Stanton Moor', *Landscape Research*, 42(6): 616–633.

Edensor, T (2020) *Stone: Stories of Urban Materiality*. Melbourne: Palgrave.

Edensor, T and Hughes, R (2019) 'Moving through a dappled world: the aesthetics of shade and shadow in place', *Social and Cultural Geography* (with Rachel Hughes), doi: 10.1080/14649365.2019.1705994

Giblett, R (2007) 'Shooting the sunburnt country, the land of sweeping plains, the rugged mountain ranges: Australian landscape and wilderness photography', *Continuum*, 21(3): 335–346.

Greenlaw, L (2006) *Between the Ears*: 'The Darkest Place in England', *BBC Radio 3*, 25 March 2006.

Howes, D (2010) *Sensual Relations: Engaging the Senses in Culture and Social Theory*. Ann Arbor: University of Michigan Press.

Ingold, T (2008) 'Bindings against boundaries: entanglements of life in an open world', *Environment and Planning A*, 40(8): 1796–1810.

Ingold, T (2010) 'Footprints through the weather-world: walking, breathing, knowing', *The Journal of the Royal Anthropological Institute*, 16: S121–S139.

Ingold, T (2016) 'Reach for the stars!: light, vision and the atmosphere', in D Gunzburg (ed.), *The Imagined Sky: Cultural Perspectives*. London: Equinox Publishing, pp. 215–233.

Kite, S (2017) *Shadow-Makers: A Cultural History of Shadows in Architecture*. London: Bloomsbury.

Lingis, A (1998) *Foreign Bodies*. Bloomington: Indiana University Press.

MacPherson, H (2009) 'The intercorporeal emergence of landscape: negotiating sight, blindness, and ideas of landscape in the British countryside', *Environment and Planning A*, 41: 1042–1054.

Miles, M (2013) 'Light, nation, and place in Australian photography', *Photography and Culture*, 6(3): 259–277.

Miles, M (2008) 'Focus on the sun: the demand for new myths of light in contemporary Australian Photography', *Australian and New Zealand Journal of Art*, 9(1–2): 220–239.

Pallasmaa, J (2012) *The Eyes of the Skin: Architecture and the Senses*. London: John Wiley and Sons.

Plummer, H (2012) *Nordic Light: Modern Scandinavian Architecture*. London: Thames and Hudson.

Simpson, P (2019) 'Elemental mobilities: atmospheres, matter and cycling amid the weather-world', *Social and Cultural Geography*, 20(8): 1050–1069.

Tredinnick, M (2013) 'The weather of who we are: an intimate essay on the weather, the self, and Australianness', *World Literature Today*, 87(1): 12–15.

Urry, J and Larsen, J (2011) *The Tourist Gaze 3.0*. London: Sage.

Wylie, J (2006) 'Depths and folds: on landscape and the gazing subject', *Environment and Planning D: Society and Space*, 24: 537–554.

Young, D (2006) 'The colours of things', in P Spyer, C Tilley, S Kuechler, W Keane (eds.), *The Handbook of Material Culture*. London: Sage, pp. 173–185.

Young, D (2011) 'Mutable things: colours as material practice in the northwest of South Australia', *Journal of the Royal Anthropological Institute*, 17(2): 356–376.

11 Foggy landscapes

Maria Borovnik and Kaya Barry

Introduction

Fog is a fascinating, always-in-motion weather phenomenon: it comes and goes, sometimes rather suddenly. Fog drifts in and blankets out landscape, over-saturating the visual. As a state of precipitation and cloud formation it can come across as intense, encapsulating the landscape and objects in an uncanny way. Fog appears to embrace the landscape and, in doing so, it temporarily immobilises humans and nonhumans and reconfigures corporeal movements, the way landscape is perceived, and the way people interact and relate socially with each other and with landscape. Fog may appear slowly, drifting in from the horizon, or it can be a sudden transition as one moves into it and is enveloped. In this chapter, we explore the different types of mobilities that are involved with fog and how these have been represented creatively. In literature and the arts, fog has cast a mysterious veil over how humans relate to and move with landscape. We are interested in fog as a climatological entity but also in the way it shapes and reshapes our relationships with landscape.

This chapter unfolds in three sections. As a first step to understanding fog, we describe the temporalities that fog encompasses in seasonal contexts or in relation to other weather conditions. We draw on two central conceptualisations of fog from Martin (2011) and Ingold (2010, 2011) to show how fog has been theorised as an intermediary between humans and landscapes, and then extend the discussion by emphasising the human and nonhuman entanglements that it produces. We also understand fog as a particular form of ever-changing cloud. For Cynthia Barnett (2015), cloud formations are more than just science, they are filled with character and mystery. Clouds, she found, could 'create moods' (2015: 2), generate an atmosphere or take on their own character. Our argument leans on these ideas and emphasises fog as more than a weather phenomenon, but as an active, yet unpredictable, agent of its own. Accordingly, we enquire into the ways in which fog alters as emerging and dissolving state of weather and our imaginations of how we move through a foggy world. We do so by drawing on conceptualisations in literature and artworks that highlight the 'mystery' of how fog reorders the mobilities that take shape within landscapes. In the final section, we argue that fog breaches human and nonhuman boundaries, and merges sensations and mobilities in a way that brings landscapes into a closer encounter with those experiencing it.

Always moving: framing fog

Climatologists define 'fog' as a form of cloud near the ground that creates conditions of low visibility: less than one kilometre (Linacre and Geerts, 1997; UK Met Office, 2018). It is perceived as particularly dense when visibility is below 40 metres (Linacre and Geerts, 1997: 153). Very small water drops in the air cause the thickness of fog. It is often confused with 'mist', which, in contrast, has a larger drop size and is consequently more visible. In a meteorological sense, 'high fog' was labelled in Abercromby's International Cloud Atlas (International Meteorological Committee, 1896: 13) as *stratus* – a low-lying cloud – and this description has stayed (Hamblyn, 2017: 63). During winter nights, fog occurs when cool air close to the surface becomes saturated with moisture (also termed 'radiation fog'), or 'when a warm front passes over an area with snow cover' (UK Met Office, 2018). Interestingly, fog appears under two almost opposite circumstances: either by cold air passing over a warm surface or by warm air floating over a cold surface (Linacre and Geerts, 1997: 156). Because of these opposite conditions, fog often forms around lakes, rivers and oceans. Sea fog often starts in early summer when the land warms up more quickly than the sea. On rainy days, fog can rise out of valleys, blown upwards by mild wind, lying along hillsides (Linacre and Geerts, 1997; Tang, 2012). Fog disappears when dry air arrives, or it may be blown off by wind, or immersed in warm sunlight.

These meteorological descriptions, and specifically their categorisation, creates the impression that fog is a movable object: depending on the conditions, fog will appear or disappear. It will actively lean or huddle across hillsides or roll over a sea or river, and it will then, in interaction with wind, air and sun, cease. This objectification may seem true if one experiences fog by gazing at it and observing the foggy landscape. Langford and Bass (2015: 3) draw our attention to the unpredictability of fog, stating that 'Fog is unpredictable. Despite advances in technology, no one ever knows for sure when it will come. Once it is here, no one knows how long it will stay'. This quote gives the impression of fog as a subject with its own will, a nonhuman entity with considerable agency. Further, when tactile senses are involved and one is surrounded by fog, one's visual perception changes and it is possible to feel as inseparably intermingled within its landscape, as being part of it. This idea of immersion is applied by Ingold (2011: 117), who suggests that 'in an open world' the objectifying boundary lines around particular weather occurrences and landscape will dissolve. In this sense, fog *is* a mobile entity with agency: rolling, cleaving, appearing and disappearing. Fog envelops landscapes, entangling humans and nonhumans within its embrace.

In this conceptualisation, both fog and the changing landscape are 'formative and transformative *processes*' in which those that are perceiving these weather landscapes are part of such phenomena in becoming (Ingold, 2011: 117–118; original emphasis). Fog and those that are swept up *in* it are part of a world that 'is inhabited is woven from the strands of their continual coming-into-being' (Ingold, 2008: 1797). It is through the feeling of and breathing with weather-scapes that living beings become part of the intermingling of solid sphere and sky (Ingold, 2007, 2011). Using the example of landscape in mist, Ingold explained that weather is

> what we perceive *in,* underwriting our very capacities to see, to hear and to touch. As the weather changes, so these capacities vary, leading us not to perceive different things, but to perceive the same things differently … it is not a figment of the imagination but the very *temperament* of being.
>
> (Ingold 2011: 130, emphasis in the original)

Ingold's intermingled view of weather and landscape takes into account more than just visual awareness, and includes other senses as part of the full enmeshment experience. Both Sørensen (2014) and Martin (2011) also foreground the importance of the value of connectivity between materiality outside a body and acoustic and tactile experiences. In Barnett's (2015) observation, weather also is related to more than the visual. She draws on Helen Keller, who remembered the 'smell of a coming storm', which she could sense 'hours before there was any sign of it visible' (Keller, 1904: 66, in Barnett, 2015: 212). Similarly, Barnett finds that the combination of acids or turpentine found in materials and plant essences from trees, conifers or tiny mosses released in the atmosphere and combined with moisture are responsible for various scents. The different types of scent are unique to the landscape and weather condition. Martin (2011: 461–462) highlights a dual process between body and environment as a form of envelopment that can be particularly and clearly sensed within fog. He explains fog as a form of gathering, a kind of bridge 'at the interstices of incoming and outgoing – it is neither body nor place, but rather acts as a gathering force, intensifying the immanent connection between the two'. With the fog's gathering, 'something is brought towards' a body, and this process of movement in itself may produce sensibilities that one may experience in the process, argues Sørensen (2014: 66). Consequently, fog not only acts as its own moving entity but it creates a relationship with landscape and those (bodies) that are involved in it.

Within such moments, however, when fog gathers together disparate actors (human and nonhuman), the 'luminescent glare' that fog creates, makes light central to experiencing its landscape (Martin, 2011: 458). This visual perception of landscape and its intermingling with light is explored by Edensor (2017), who emphasises that landscape itself is in constant motion, loaded with energy and always changing. Light and visual experience, together with other senses such as touch or smell, interact with each other and are entangled in the experiences and imaginations of landscape (see Chapter 10). Edensor (2017: 617) contends that 'landscapes seethe with multiple rhythms and temporalities as elements within it incessantly emerge, decay, die and transform'. He agrees with Ingold (2011) that this liveliness, this vitality of landscape, can be felt with all senses, and wind and water are part of this. Yet, light must be given specific attention as light effects, shades and levels play a role in how landscape resonates emotionally. It is this sensing of fog landscapes and the particular resultant perceptions and affects upon which we focus in this chapter.

Depending on the particular weather conditions and the number of particles in the air, the light levels of fog vary from bright and luminesce to thick and dim. Martin (2011) elaborates on the loss of horizon under dense fog. Intriguingly, the void or boundless emptiness that Ingold (2008: 1799) identifies when staring at a blue sky is perhaps echoed by the full thickness in fog. The density and opacity experienced in fog is so close to a person or landscape that both may feel its embrace like clothing.

Fog landscapes in literature

Central to our conceptualisation of fog is the ability to gather (Martin, 2011) and intermesh (Ingold, 2010, 2011) humans within landscape. Although these two central theorisations have informed our understanding of fog, we want to emphasise the agency that this weather phenomenon holds, as a force that entangles humans with the nonhuman realm through its mobility. Weather is often used to set the mood or atmosphere in literature, creative writing and the arts to capture the feelings and sensations of changes in how humans relate to, and move within, landscapes. Thinking through a geographical lens, the notion of atmospheres and often intangible qualities escape easy representation (Anderson, 2014; Vannini, 2015), yet are strikingly present in a rich assortment of creative expressions. While a novel, poem, artwork or other form of creative expression may be at first understood as 'representing' landscape, it is the force of the atmospheres that they express that fascinates us. Thinking through a non-representational approach (Anderson, 2014; Boyd and Edwardes, 2019) allows for a reading and reflection on creative expressions of weather that constantly seem to escape our grasp. Moving beyond purely representational (or meteorological accounts) of weather, in some sense, seeks to entertain the 'what-if' speculation of the interactions we have with 'a myriad of other sensations, [that] inform and alter an individual's perception and attentions within the world' (Barry, 2019: 205). How we 'converse' with landscape (Benediktsson and Lund, 2010) often goes beyond a singular sensation, opening us to the realm of multisensory forms of attention. We explore creative literature that uses fog as an active element in the landscape for mobilising different affects, impressions and experiences of this unpredictable, fast-moving, yet, often, lingering weather phenomenon. We look at two different, almost opposite kinds of descriptions of fog landscapes in literature: those that express wonder and magic, and those that conversely articulate the notion of fog as lifeless, melancholic and bearing danger. In both descriptions, fog appears as a being, an entity that does something to landscape, a creator.

In Carl Sandburg's *Chicago Poems* (1916), fog is described as approaching on 'little cat feet', where it is a lovely creature, filled with mystery, taking the form of a cat that 'silently haunches and then moves on'. This fog–cat is a free creature, overlooking the 'harbor and city' landscape of 1916 Chicago. Landscape bathed in fog is also perceived as very quiet by David Langford in his preface to a photographic book titled *Fog at Hillingdon* – 'so quiet that your thoughts seem to be amplified. Fog even seems to quieten the clanging, banging chaos of cities' (Langford and Bass, 2015). The photographs in the book inspire us to consider how fog does not mute but rather enhances colours in the landscape. The predominant grey foregrounds the colours of an autumn landscape, and the photos that are presented in the book give evidence for this observation, as the quote below elaborates:

> Fog has its own personality. … Sometimes fog hovers only over the hilltops, and the valleys and lowlands are virtually clear. At other times it is the complete opposite. … Some fog moves horizontally across the landscape. Then, there is the wispy fog that rises like smoke on the water. Sometimes fog freezes as it is suspended in the air, and clings to grasses, brush, and trees.
> (Langford and Bass, 2015: iii)

But mostly, the photographs and contributing authors emphasise the magic of fog, bringing peace and clarity to one's mind, as expressed by Andrew Sansom in his foreword:

> And so, as I look out my window on a frigid day in January, my mind and my soul are drawn to the enchantment and elegance of the serpentine river of fog below me. My soul is moved by the solemnity and grace of the scene, and my mind is certain that through the fog there is clarity.
> (In Langford and Bass, 2015: i)

These affects of enchantment, solemnity, being moved and finding clarity contrast with what can be found in many examples of fiction that use fog as a medium to underline heaviness, depression or secrecy. For example, in Cormac McCarthy's *The Road* (2006), fog is hardly overtly mentioned but is ever present in the way in which the landscape is described as blanketed out, profoundly affecting people:

> They walked out and sat on a bench and looked out over the valley where the land rolled away into the gritty fog. A lake down there. Cold and grey and heavy in the scavenged bowl of the countryside.
> (2006: 10)

McCormack's cold, grey and heavy description of the landscape, which was made so by fog, is described here as 'gritty', coarse, and relentlessly produces the hopeless and endlessly heavy atmosphere in the post-apocalyptic scenery in which a man and his young son continue to walk in a persistent opacity. Fog is not explicitly mentioned again, and yet the entire book hazes its landscape; in its never-changing grey and dull appearance, this landscape is estranged and uncanny (Nancy, 2005).

Other examples of popular fiction that employ fog as an active agent to set a certain tone of landscape include David Guterson's (1995) *Snow Falling on Cedars*. Fog plays a significant role in the book's key event: a fisherman is killed during a completely immobilising, stunning, opaque night out in the bay. Fog is subsequently also used as descriptor throughout the book, which depicts the court case surrounding the death, and several complex relationships that are interwoven within the community. Where cold and heavy fog had weighed in the apocalyptic McCarthy novel, in Guterson's work it accompanies the ever-lurking, dangerous possibilities of injustice. The thick disorienting, immobilising and yet ethereal fog is contrasted to how snow may act as a veil behind which hidden pleasures can be experienced. Fog, however, can also function as a veil or cover, where the world is blurred and put on hold, as the short story *Fog* by Kent Nelson (2014) describes. In this story, a place is 'shrouded in fog' (446), where instead of 'seeing colors' the 'effect of greyness' is that people now 'cope with the muting of the background' (451), a background that entails more than landscape. This blurred or muted landscape includes an inner psychological scape in which destruction and self-destruction grows, and where someone can feel like a 'caged animal' (2014: 452).

So, fog can come across as both beautifying landscape, as in the *Fog at Hillingdon* descriptions, supporting clarity of mind either outwardly or inwardly, or as weighting on mind and landscape, symbolising unbearable hopelessness or the dangerous unpredictability of life. Old English stories use fog to veil looming danger. During a cold winter evening, the moors of *Wuthering Heights* in Emily Brontë's novel (1847), which are snowed under and covered by fog, contain a depression that anticipates a dangerous narrative that will follow, filled with unfortunate twists and varying shades of oppression. More exaggeratedly, in the criminal story of the *Hound of the Baskervilles* by Arthur Conan Doyle (1901: 103), fog is used to create a gloomy atmosphere, stoking anticipation that something uncanny will soon occur:

> A dull and foggy day, with a drizzle of rain. The house is banked in with rolling clouds, which rise now and then to show the dreary curves of the moor ... and the distant boulders gleaming ... [creating] melancholy outside and in.

In Doyle's description, fog has become a painter, capturing landscape as dreary or gleaming. In using shades of light and dark, the moorland appears melancholic within fog, affecting the feeling of those inside. For instance, Watson feels conscious 'of a weight at my heart and a feeling of impending danger – ever present, which is the more terrible because I am unable to define it' (Doyle, 1901/2007: 103).

A wealth of literature refers to the pea-souper fog of early modern London. Rapid industrial expansion in Europe and the introduction of coal to domestic fires had created problems of pollution in many cities, and, in particular, those that were predisposed to fog such as London, located in the Thames Basin and surrounded by hills. Smog, or smoke fog, is typical of damp, polluted urban climates when low-level inversions meet air pollution (Linacre and Geerts, 1997: 157). The UK Met Office (2018) notes that the thickest fog appears in industrial areas, where polluted particles help water drops to grow. Already by medieval times, a polluted, smoky and foggy climate was observed in London, especially after the introduction of sea coal (Corton, 2015: 1f, drawing on Brimblecombe, 1987: 8 and 30). Corton (2015: 1) traces the typical yellow, pea-souper London fog to the period between the 1840s and the 1880s. Even so, air pollution continued to be an issue in London. The London fog in 1952 caused fatalities and illnesses (Corton, 2015; WHO, 2013). As a consequence, in the 1960s, the UK Clean Air Act was introduced, leading to a decline in pollution.

Much literature and art has drawn on London fog; Charles Dickinson's late 19th-century novels (for example, *Bleak House* and *Our Mutual Friend*) are packed with fog analogies, using the greyness or blackness of fog as signs of corruption and greed encroaching on the urban. Henry Green (1947) uses fog as symbolising a retreat into insecurity and leading to a standstill, while Sam Selvon (1956) deploys fog to emphasise London's urban isolation and how fog delays people's movement. In Thomas Stearns Eliot's *The Love Song of J. Alfred Prufrock*, originally published in 1915, just one year after he had moved to London, 'yellow' fog is referred to as a pet, likely a cat (similar to Sandburg) that licks, lingers, falls, slips and curls, and 'rubs its muzzle on the window panes' (Elliot, 2015: 9). Different from Sandburg's fog-cat, this creature is scruffy, yellow and smoky, reflecting a sooty, draining environment.

What these examples have in common is that fog is unreliable in its appearance; it is temporal, ephemeral, takes on cloaks of different colours and is unpredictable – one never knows when or whether it may stay or leave. The loss of orientation or horizon in these different descriptions and narratives impose momentary (or lasting) outward disorientation or immobility. And yet, this same disorientation can also enforce inner change; one is stunned by the anticipation of danger or the depressive, tangible encroachment of lightlessness, or conversely, moved by the sudden clarity behind the grey. In this way, fog connects between the material outside and our human bodies and senses (Martin, 2011; Sørensen, 2014). It actively functions in what Ingold (2011) identifies as formative processes. These may involve the multisensory experience of interacting with the greyness, dampness, and enclosure of fog, and a sense that this chimes with one's inner being.

Immersive artistic responses to fog

Fog – as cloud, aesthetic and experience – while featuring prominently in literature, has also been explored through creative artworks that expand on aspects of illusion and the immersive states that fog can produce. In the following section, we survey examples from contemporary art to show how artists have used fog as a particular weather phenomenon to explore notions of embodiment, liminality and human–environment encounters.

The Japanese artist Fujiko Nakaya has been working with fog in her large-scale environmental installations since the early 1970s. Interested in the relationships between humans and nature, one of her best-known works that is still active today, *Foggy Wake in a Desert* (1982), is a permanent installation in the 'Sculpture Garden', a lush rainforest area that contrasts against the dryer scrub landscape on the perimeter of the National Gallery of Australia, Canberra. A series of mist-emitting sprinklers are activated at a scheduled time each day that create a large fog drifting across a shallow pond and undergrowth. Her 1976 installation at the Second Biennale of Sydney, *Fog Sculpture #94768: Earth Talk*, received international acclaim and led to decades of fog exhibitions at large cultural institutions and festivals.

Nakaya's interest in fog began in the early 1970s when she collaborated with scientists at Kyoto University to conduct a decade-long observation of the impacts of 'inserting a one square kilometer fog in desert conditions' and recording the results (National Gallery of Australia, 2020). Other fog artworks are installed across existing architectures and surfaces, including walking paths, platforms and even along bridges. She describes fog as 'a medium for the transmission of light and shadow, much like video', in which she became interested through thinking about 'decomposition' or 'the process of decaying' (Guggenheim Bilbao, 2020).

Another prominent example is the infamous artwork series of Berndnault Smilde, titled *Nimbus* (2012–2017), in which he created small clouds in a gallery (see Slobig, 2015; Smilde, n.d.). Existing just long enough to be photographed, these 'clouds' were created with fog machines and had a short temporality; yet due to extensive study and planning, they took the shape of 'nimbus' clouds. Although these artworks dispersed almost instantaneously, some lasting around ten seconds (Slobig, 2015), Smilde's *Nimbus* series attracted international media attention and continues to be influential to scholars and scientists due to the technical proficiency of the created nimbus cloud formations. While the foggy clouds exist only in photographic documentation, Smilde purposefully created these instances in empty architectural spaces such as cathedral interiors, large halls and even at the top of escalators and stairwells with high ceilings above. The absence of any human presence in the photographs evokes a spiritual, ephemeral appearance, with the clouds maintaining an allure and illusion that shapes the perception of the architecture in particular ways.

Another example of exploring fog-like weather is found in the artists Diller and Scofidio's 2002 installation, *Blur Building*. In this large, site-specific installation, Diller and Scofidio explore how bodies interact with and respond to changes in weather that co-constitute to the environmental surrounds. Composed of a complex series of engineered fog machines, lighting, a weather station and ramps for pedestrians, the installation invites visitors to enter the artwork through an immersive foggy space. Visitors wear 'bio' lab coats that are equipped with motion sensors that transmit information about their movements that influence the shape of the fog and lighting. Diller and Scoffido term this project an 'architecture of atmosphere' rather than an artwork, in which up to 400 participants can interact and collectively alter the atmospheric conditions within which they are immersed.

There is also Danish-Icelandic artist Olafur Eliasson's famous *Weather Project* installation at the Tate Modern Turbine Hall in 2003 (and retrospective in 2019, see Eliasson, 2019), in which a light foggy mist was created to fill the immense hall. High above the visitors was a giant 'yellow orb like a dark winter sun' (Grynsztejn, 2007: 11). Since this 2003 installation, Eliasson has experimented with weather phenomena and colour in many large-scale installations, such as the *Fog Assembly* (2016), *Feelings Are Facts* (2010), and *Your Blind Movement* (2010), in which red, green and blue lights oversaturate a room filled with a fine fog. In the two colour-fog installations, visitors were invited (and encouraged) to move around the room, riding bicycles, running and jumping.

At the core of Eliasson's body of work is making felt the relationships between bodily experience, immersive atmospheres and the qualities of perception that are brought to the foreground through these liminal environments. In the exhibition catalogue for *Fog Assembly*, Madeline Grynsztejn (2007: 15) describes his works as exploring the 'substance of experience', which 'is not prescribed but rather corporeally *enacted* from moment to moment, a realisation that is subsequently available for transposition to the world at large'. Eliasson's work includes collaborations with notable scholars such as Gernot Bohme and Doreen Massey, exploring the relationships of spatiality and perception, aesthetics and atmospheric conditions. He states:

> we are often numb to the atmospheres that surround us. Here, architectural detail and artistic intervention can make people more aware of an already existing atmosphere. That is, materiality can actually make atmospheres explicit – it can draw your attention and amplify your sensitivity to a particular atmosphere.
>
> (Eliasson, in Borch, 2014: 95)

These are just a few examples of how contemporary artists working across architecture, installation, arts practice and geographical inquiry have explored fog and landscape through sensuous and practice-based methods. Adding to the accounts in literature and creative writing that we explored earlier, these arts examples bring to the surface how fog's immersive qualities highlight the range of sensations – often beyond the visual – that fog (and other weather conditions) produces in the human body. While Diller and Scofidio's installation with the bio lab coats invites the participant to be active in the configuration of the installation's landscape, Smilde's *Nimbus* photographs reveal how fragile and momentary such immersive weather experiences may be. Eliasson's installations epitomise the sensory bias we have towards the visual, and by using thick, heavy clouds to produce foggy landscapes, urges participants to recalibrate their sensations to the weather. Enveloping and immersing, drifting and encroaching, the foggy installations described here all seek to bring the visitor or perceiver of the artwork into new relationships by using their body as a barometer.

Conclusion

In this chapter, we have built on existing scholarly work on fog (Ingold, 2010, 2011; Martin, 2011) that draws eloquently on those contemplative moments of moving-with landscape. Tracing a range of examples from literature and the arts, we have shown the variations in how fog has been used as a weather phenomenon to significant alter how humans perceive and experience the tonalities and textures of landscapes.

The inspiration for this chapter emerged from our fascination with how fog has been employed in creative expressions, as an active agent that mobilises and transforms a landscape, and is not merely conceived as a passive phenomenon. If we consider the relations between humans and fog, then fog can seem to be an unpredictable 'being' or actor, which lies across space or clothes landscapes and, by doing so, enmeshes humans within these layers. While fog and cloudy atmospheres have been described rigorously by Ingold and Martin, we extend their largely personal ethnographies by exploring the importance of fog in creative forms of expression, in novels, poems, photography and contemporary art. When mobilised in these creative works, fog becomes evident as a nonhuman performer in the landscape, possessing an agency and unpredictability that fascinates human attention. The multitude of ways that fog has been captured and represented to express feelings, emotions and atmospheres that further entangle humans and landscape, shows how 'passive' weather phenomenon is often accentuated when we engage with it in creative forms. As a fickle, mysterious, ominous force or that which spectacularly alters human orientation, the foggy clouds near the ground force us to relate and move with landscape in unpredictable ways.

References

Anderson, B (2014). *Encountering Affect: Capacities, Apparatuses, Conditions.* Farnham: Ashgate.
Barnett, C (2015). *Rain. A Natural and Cultural History.* New York: Broadway Books.

Barry, K (2019). 'The echo of communal space: more-than-representational accounts of tourist encounters in hostel accommodation', in Doughty, K, Duffy, M, Harada, T (eds.), *Sounding Places: More-than-representational Geographies of Sound and Music*. London: Edward-Elgar. pp. 202–212.

Benediktsson, K and Lund, K (2010). *Conversations with Landscape*. Farnham: Ashgate.

Borch, C. (Ed.). (2014). *Architectural atmospheres: On the experience and politics of architecture*. Basel: Kirkhauser.

Boyd, C and Edwardes, C (eds.) (2019). *Non-Representational Theory and the Creative Arts*. London: Palgrave.

Brimblecombe, P (1987). *The Big Smoke: A History of Air Pollution in London since Medieval Times*. London: Methuen.

Brontë, E (1847/2012). *Wuthering Heights*. London: Penguin.

Corton, Ch L (2015). *London Fog*. Cambridge, MA: Harvard University Press.

Doyle, AC (1901/2007). *The Hound of the Baskervilles*. London: Penguin; originally London: George Newnes.

Edensor, T (2017). *From Dark to Light. Daylight, Illumination, and Gloom*. Minneapolis: University of Minnesota Press.

Eliasson, O (2019). Fog assembly, 2016 [website]. Available at: https://olafureliasson.net/archive/artwork/WEK110139/fog-assembly

Elliot, TS (2015). *The Love Song of J. Alfred Prufrock and Other Works*. Boston: Squid Ink.

Green, H. (1947). *Party Going*. London: Hogarth.

Grynsztejn, M (2007). '(Y)our entanglements: Olafur Eliasson, the museum, and consumer culture', in Grynsztejn, M (ed.), *Take Your Time: Olafur Eliasson*. New York: Thames and Hudson, pp. 11–32.

Guggenheim Bilbao (2020). Fog Sculpture #08025 (F.O.G.). Fujiko Nakaya. Guggenheim Bilbao. Available at: https://www.guggenheim-bilbao.eus/en/the-collection/works/fog-sculpture-08025-f-o-g

Guterson, D (1995). *Snow Falling on Cedars*. London: Bloomsbury.

Hamblyn, R (2017). *Clouds: Nature and Culture*. London: Reaktion Books.

Ingold, T (2007). Earth, sky, wind, and weather. *Journal of the Royal Anthropological Institute*, (N.S.): S19–S38.

Ingold, T (2008). Bindings against boundaries: entanglements of life in an open world. *Environment and Planning A*, 40: 1796–1810.

Ingold, T (2010). Footprints through the weather-world: walking, breathing, knowing. *Journal of the Royal Anthropological Institute*, (N.S.): S121–S139.

Ingold, T (2011). *Being Alive: Essays on Movement, Knowledge and Description*. London: Routledge.

International Meteorological Committee (1896). *Atlas International des Nuages/International Cloud Atlas/Internationaler-Wolken-Atlas*. Paris: Gauthier-Villars.

Keller, H (1904). *The World I Live In*. New York: The Century Company.

Langford, D and Bass, R (2015). *Fog at Hillingdon*. Austin: Texas A&M University Press.

Linacre, E and Geerts, B (1997). *Climates and Weather Explained*. London: Routledge.

Martin, C (2011). Fog-bound: aerial space and the elemental entanglements of body-*with*-world. *Environment and Planning D: Society and Space*, 29: 454–468.

McCarthy, C (2006). *The Road.* New York: Vintage.

Nancy, JL (2005). *The Ground of the Image.* New York: Fordham University Press.

National Gallery of Australia (2020). Fujiko Nakaya: foggy wake in a desert: an ecosphere. *National Gallery of Australia.* Available at: https://cs.nga.gov.au/detail.cfm?irn=48360

Nelson, K (2014). Fog. *Antioch Review,* 72(3): 443–456.

Sandburg, C (1916). *Chicago Poems.* New York: Henry Holt and Co.

Seldon, S. (1956/1984). *The Lonely Londoners.* Harlow, Essex: Longman Drumbeat.

Slobig, Z (2015). How this artist makes perfect clouds indoors. *Wired,* 6 May 2015. Available at: https://www.wired.com/2015/06/berdnaut-smilde-nimbus/

Smilde, B (n.d.). Nimbus. Available at: http://www.berndnaut.nl/works/nimbus/

Sørensen, TF (2014). More than a feeling: towards an archaeology of atmosphere. *Emotion, Space and Society* 15: 64–73.

UK Met Office (2018). *What Is Fog?* Available at: https://www.metoffice.gov.uk/learning/clouds/fog

Tang, Y (2012). The effect of variable sea surface temperature on forecasting sea fog and sea breezes: A case study. *Journal of Applied Meteorology and Climatology,* 51: 986–990.

Vannini, P (ed.) (2015). *Non-Representational Methodologies: Re-envisioning Research.* New York: Routledge.

WHO (2013). *World Health Report 2013: Research for Universal Health Coverage.* World Health Organization: Geneva. Available at: http://www.who.int/whr/en/

12 Sensing bushfire

Exploring shifting perspectives as hazard moves through the landscape

Katharine Haynes, Matalena Tofa and Joshua Whittaker

Introduction

This research was conducted on Yuin Country. We acknowledge the Elders of Yuin Nation past, present and emerging and their work of caring for Country and reducing fire risks. When people recount their experience of living through a meteorological or geological hazard, such as a flood, bushfire or cyclone, it is the seeing, feeling, smelling and hearing of distinct elements that often dominate post-disaster dialogues, and remain the core vivid memories long after the event. For example, the sounds of Cyclone Tracy, a Category 5 cyclone that made landfall on Christmas eve 1974 in Darwin, the northernmost city of Australia, were strongly remembered almost 30 years later by those who had lived through the experience: 'you could hear all the windows smashing, just like someone was hitting it. Then it was roaring, it was really loud, it was deafening' (Haynes et al., 2011: 36). Similarly, people often describe the sounds they encounter during floods:

> The noise of the water coming across here, was deafening. It was terrifying. It was like jumbo jets on a runway... I could hear everything under the house bashing around... I can still feel that absolute terror of listening to the river. That noise, that was unbelievable. It's lodged somewhere deep in my soul, that noise.
>
> (Tofa et al., 2017: 66)

For some residents, it was the smell of the flood that remained: 'the smell is horrific. It will never leave my nostrils. The other thing is [the flood] covers everything in mud and everything dies and then that rots...' (Tofa et al., 2017: 66). Often it is the absence or changed ability to sense during a disaster that dominates. The Australian bushfires of 7 February 2009 that killed 173 people, known as Black Saturday, turned 'day to night' making well-known gardens and communities suddenly impossible to navigate: 'it got darker and darker until it was totally pitch black' (Whittaker et al., 2017: 123). Yet, despite their likely influence on decision-making, (im)mobilities and sense of place, little scholarly work has explored people's sensuous experiences of hazard events.

Rodaway (1994: 3) states that our senses provide mediated information about the world around us and are thus 'the ground base on which a wider geographical understanding can be constructed'. We borrow, as Rodaway (1994) does from McLuhan (1962), an interpretation of the senses as both a medium for receiving information and also a message. Importantly, the senses do not inertly receive environmental information but actively decode and order the knowledge received (Rodaway, 1994); put differently, the 'sensorium' of the five senses (smelling, hearing, feeling, seeing and tasting) filters information about the surrounding environment into meaning (Howes, 2019). We extend this meaning to include current messages of the environment, as well as those removed in time and space. For example, a smell or scent may provide an immediate message about the environment, but can also trigger strong emotional memories of past experiences.

Following Lund (2005) and Paterson (2009), feeling is interpreted not to simply mean 'touch' but to encompass immediate bodily experiences, such as pressure on the skin or sensations of heat, cold or pain, which are accompanied by feelings throughout and within the body. Within our interpretation of 'felt' we use Paterson's (2009: 768) 'somatic senses' to acknowledge the intimate interaction between the external and internal senses. Paterson (2009), further, draws the important distinction between 'sensations' that are simply responses through nerves and sense-system clusters and 'sensuous dispositions' that are constructed through sociohistorical and contextual processes.

Senses, then, do not simply provide an unmediated 'truth' about the environment, surroundings or place and several authors have demonstrated the importance of considering the 'culturisation' of senses (Howes, 2019). Cultures can in themselves be defined as 'different ways of sensing' the world (Howes, 1991: 8), and, in this way, the senses are not only means of inquiry but also objects of study in themselves (Howes, 2019). As Howes (2019: 22) puts it, the senses 'are not simply information seekers or drones; rather, they are culture bearers and therefore always subject to moral regulation'. Each culture, thus, must be understood on its own sensory terms, and differences in sensory values are also likely within each society, persons and groups (Classen, 1997). Indeed, senses and sensations are gendered, racialised and structured by social class (Bourdieu, 1987: cited respectively in Howes 2019: 22; Classen, 1998; Stoever, 2016). Interpreting and understanding messages decoded and received through the senses is culturally shaped, and generative of individually different ways of smelling, hearing and seeing to provide different understandings (Howes, 2019; Howes and Classen, 2013). These differences are consequential when considering responses to extreme weather events, where divergent interpretations of sensory experiences can prompt a range of responses which may increase risks or enhance public safety.

We also look to the literature on emotional and cultural geographies, closely tangled with the sensory, and important to our sense of place. As noted by Davidson and Milligan (2004: 524), our emotions have 'tangible effects on our surroundings and can shape the very nature and experience of our being-in-the-world. Emotions can clearly alter the way the world is for us, affecting our sense of time as well as space'. Similarly, relationships with place are always temporary and changing (Bartos, 2013), and 'place' can thus be understood as a process that is not finished and always in a state of 'becoming' (Pred, 1984). This porosity of boundaries between outer and inner worlds, and the continual remaking of places and relationships is suggestive of the importance of sensory and emotional experiences of extreme weather events. Cultural geographers provide some useful insights to the interactions of bodies in and with places. Wylie (2005: 239), for instance, provides a vivid account of the intertwining of body, geomorphology and vegetation while walking along a densely overgrown narrow coastal path 'limbs and lungs working hard in a haptic, step-by-step engagement with nature-matter'. He discusses how the boundary between 'landscape, subject and object, could become "soluble and osmotic"' (239). In the context of navigating hazards, although, this embodied engagement may open to experiences of discomfort or impaired/disrupted senses. Cultural geographical accounts of pain and discomfort encountered within the landscape and their influences on (im)mobilities are rare. A noteworthy example comes from an account of Mawson's 1912 polar expedition (Yusoff, 2007: 227) where the explorers recount how

> our eyes were more or less out of action…you get a sense of touch which nothing else except bare feet could give you. Thus, we could feel every small variation in surface… With no light or landmarks to guide them they 'travelled on by the ear, and the feel of the snow under our feet.

Indeed, navigating the hazard and landscape typically invokes feelings of disorientation, scrambling and 'stuckness' (Bissell and Gorman-Murray, 2019; Straughan et al., 2020). Such disorientations, too, are 'palpably felt' as an encounter with distance and limits, and, increasingly, with disconnection (Bissell and Gorman-Murray, 2019). During a hazard event, information on the unfolding disaster, what to do and where to go are important aspects, often influencing people's decision-making. Indeed, developments in mobile technologies have arguably reshaped the mobilities of communication and information about hazards, weather and travel routes. This boom in readily available real-time information during emergencies has been dramatic, with a proliferation of applications, websites and social media commentary (Houston et al., 2015). However, coverage is not equally distributed. Spaces exist where mobile phone coverage remains poor, and during disasters, communication networks are often disrupted. Commentary following almost every disaster includes vociferation that official messaging was poor, not detailed enough or not received at all (Dash and Gladwin, 2007; Haynes et al., 2009). In instances where official communications have been disrupted, or inadequate, people have to rely on their own senses and sense-making to navigate the hazard. For example, people often first become aware of a possible bushfire threat through smelling or seeing smoke (Whittaker et al., 2013). Even when more formal messages are received, people often need to verify the scale of the threat themselves by actually witnessing physical evidence of the bushfire (Whittaker et al., 2013; Whittaker et al., 2020), or flood (Anzai and Kazama, 2018; Haynes et al., 2018). It is often through their sensory engagement that people experience disorientation and reorientation, and make their final decisions on what to do.

This chapter explores how senses have mediated the geographical experiences, (im)mobilities and sense of place of those who experienced a bushfire disaster on the south coast of New South Wales, Australia, in 2018. Following the example of Rodaway (1994), only smell, sound, feel and sight will be considered as they are the most appropriate in terms of understanding geographical experiences of hazard and risk. While sight and sound have been well considered in cultural geography, the 'lower' senses of feeling and smelling have not (Howes and Classen, 2013: 3; Paterson, 2009). As Howes and Classen (2013: 5) note, sensations reinforce and play off each other and can also contradict each other as 'they are an interactive web of experience'. Therefore, for this study, senses will generally be dealt with together, although at different times one sense will take prominence over another as changes in light, chemicals, temperature and sounds drift and roar through the changing hazardscape.

The Reedy Swamp Bushfire

On Sunday, 18th March 2018, a bushfire impacted on the communities of Reedy Swamp and Tathra on the New South Wales (NSW) south coast, approximately 450 km south of Sydney. The regional seaside town of Tathra is a popular tourist destination and has a resident population of approximately 1,700. The town is mostly comprised of freestanding houses on residential blocks, with a smaller number of apartments and duplex houses, as well as caravan parks, hotels and motels to accommodate tourists, and cafes, restaurants and other attractions. Reedy Swamp lies 5–10 km northwest of Tathra, on the northern side of the Bega River. It has a population of around 80 residents living on small acreages and rural residential blocks. Many of these properties are located within the forest and are accessed by unsealed roads that are flanked by vegetation.

The Reedy Swamp Fire is believed to have been caused by the failure of electrical infrastructure shortly after midday on 18th March. By 12.30 pm, the temperature had reached 37°C, relative humidity was 18% and wind was north-westerly and gusting up to 72 km/h. Although this fire fell within the bushfire danger period (1 October–31March), it was considered late in the season for such a severe and destructive fire. The fire burned in the vicinity of Reedy Swamp before embers carried the fire over the Bega River and into Tathra. The fire destroyed 65 homes and 35 caravans/cabins, and damaged a further 48 homes. Approximately 700 residents and an unknown number of tourists and visitors to Tathra were displaced on the day. Fortunately, no human lives were lost.

We were commissioned by the NSW Rural Fire Service (NSWRFS) to investigate how people in the community responded to this extreme weather event (Whittaker et al., 2018). Using semi-structured interviews, we interviewed 120 residents in Reedy Swamp and Tathra, asking about their experiences of the fire. The interviews were conducted 8–12 weeks following the fires[1]. The initial research by Whittaker et al. (2018) identified two key factors that significantly influenced how the event unfolded in Tathra. First, many people were unaware that embers can carry fire from 'the bush' into urban or suburban settings. This meant that many people in Tathra did not think they were at risk from bushfires and did not have a plan for how they would respond. Second, the loss of electricity, mobile phone reception and issues relating to the broadcast of emergency information into the local area impeded the delivery of warnings, information and advice. This meant that most people were largely reliant on their senses and those of their neighbours.

Exploring the mobilities of people and boundaries

In this section, we explore the bushfire as it was narrated by residents – beginning with how they became aware of the fire, how they responded, their reflections on their decisions, and their sense of the longer-term implications of the fire in the days and months following the fire.

Becoming aware

Without official warnings, people first became aware of the fire through a range of sensory engagements. The majority were first alerted by the smell or the visual sighting of smoke, others by the power outage and the strong wind, and some by the eerie atmospheric conditions that day. These first indications of danger led the majority to seek further information from neighbours, or to verify the threat themselves by going to a vantage point where they could better see the scale of the fire and the direction of travel:

> Both of us were sitting in for a midday snooze. Our son came in and asked if we had a fire going? I answered no. Then he said, "I think I can smell smoke." I went outside with him, I could also smell some smoke. The distinctive smell of burning eucalyptus, bushfire smell…

One resident discussed how they were having a very normal day: 'kids were playing on the water slides and the power went out, and then we could see smoke'. They went on to discuss how this led them to drive to the river mouth to get a better idea of the danger the fire posed. They also noted how even earlier in the day the atmospheric conditions were unsettling and signalled a sense of danger: 'It was really eerie. There was even nippers [surf lifesaving club for children] down on the beach and quite a lot of the parents down there were saying how it just felt weird on the beach that day'.

A minority discussed how they were anticipating danger and monitoring fire weather conditions more formally:

> I saw the temperatures shoot up to 38 and a half, those strong westerly winds arrived and then I looked at the humidity and the humidity was about 31% and I thought, well this is bad fire conditions, and I looked over to the west and I saw all the smoke coming.

Once aware of the threat, people discussed a number of sensory cues of the changing weather that led them to take action. The senses are culturally shaped (Howes, 2019; Howes and Classen, 2013) and therefore people's interpretations of what they were seeing, hearing and feeling, directly influenced and constrained what people did. People made a decision, or were forced into an action of evacuating, sheltering or defending their properties. For many this was challenging as the landscape was suddenly very different. The fire had transformed a familiar place into one that was unfamiliar and uncertain; this scrambling and disorientation was 'felt', as Bissell and Gorman-Murray (2019) stated, as an awareness of distance and limits, but also as an awareness of danger or risk, and a need to respond.

Cues that led many to evacuate included the thickening of the smoke and having difficulty breathing, seeing the smoke change colour, or the sudden darkness when the sun was blocked by smoke. For others it was seeing approaching flames or increasing spot fires and falling embers. Residents also discussed the actions of others, such as seeing neighbours leave and the general heightened sense of action – the sirens, helicopters, mayhem. For many, evacuating was not a planned response to fire, as is recommended by emergency services, but rather a bodily reaction to what their senses were interpreting:

> In terms of when do you go, when do you go? You go when you go, you know? I guess the two things that was like, "Okay, it's time to go," was when the sun was blocked out and when I saw that white smoke. I was like, "It's too close. I have to go." And then when I saw everyone else had probably left a lot earlier than me.

Another person explained:

> The smoke got really, really thick and hard to breathe, and it was just so hot. If you can't breathe, you gotta go.

Similarly, another recounted:

> The smoke started getting thicker and thicker and you could actually see some of the flames, it was just crazy, the most craziest thing I ever seen, and then the wind was blowing that strong and there was embers and stuff going everywhere and sirens, and there was just mayhem and everyone was running around and trying to get out.

These initial sensory engagements with the weather conditions, mostly incidental rather than deliberate, were how residents came to understand there was a threat and a need to respond. While it is well-known that it is often the more distal smell and sight of smoke that provide cues to get further information, the important role of the other senses, especially the feel of the strong wind and falling embers, in verifying the risk, particularly in the absence of any official information, is less well understood in emergency management. These sensory experiences arguably work as a form of confirmation, 'that it is there, that it feels like this, that we are here to experience it, that our eyes do not deceive us' (Hetherington, 2003: 1941). The 'sensorium' of the senses can therefore provide a more proximal way of knowing risk (Howes, 2019).

Navigating the hazardscape

People who evacuated by car faced the challenge of very poor visibility, with the prospect of meeting the fire, other vehicles, animals and falling trees a real possibility: 'We went through very thick smoke with zero visibility... We didn't know what was on the other side. We didn't know ... We couldn't see'. In many ways, then, the residents of Tathra became blind and utilised memory, feeling, hearing and speaking to others to make sense and find their way. This can be understood as both an intercorporeal and multitemporal experience, 'a composite of external and internal stimuli from present and post moments' (Macpherson, 2009: 1049). In this sense, navigating the hazardscape simultaneously draws on and brings together sensory experiences and memories.

All residents suffered in the smoky conditions, with the thick acrid smoke not only obscuring vision but also irritating eyes and making it difficult to breathe. 'I had a bandanna around my face so that I could breathe. I couldn't see, my eyes were stinging, they were all tearing'. The smoke coupled with the strong wind and heat made defending property extremely difficult, particularly as so many residents were unprepared and did not have appropriate clothing, face masks or eyewear (Whittaker et al., 2020). Residents described how the wind literally blew them off their feet and they could feel the heat run down the backs of their throats. The descriptions of defence are raw and provide a graphic sensory account:

> The garden started to catch fire. It was on the roof but the wind was blowing the hose up down and into the street more... it was unbelievable. There

was no air to breath. The smoke was so thick, you couldn't see across the road… it actually blew me off me feet just out there near that garden. And there was no air to breath, that was the thing I found difficult.

Other residents described how they felt as the horrific situation approached:

Me and a few other neighbors stayed, and yeah, we just got our garden hoses together as much as we could… it was just like flame balls was just hitting this tea tree and just exploding… the wind was just horrific. Armageddon… you couldn't even hardly stand up at some stages… embers flying around in the heat of it. You could even feel them. At one stage there I was at the corner and the flames were coming back up near the road. I could feel the heat… you could feel it run down the back of your throat, and the smoke, it was hard to breathe and the heat was horrific.

Here, the permeability of the body to the surrounding elements is a really useful framing for those who lived through the Tathra bushfires. Their responses were largely unplanned and raw engagements with 'nature-matter' (Wylie, 2005), literally feeling the smoke and heat inside and outside their bodies. Within this new space, people also discussed how they lost track of time due to the adrenaline and the continual effort of responding to the fire:

Initially I was totally confused because you had no idea of time. None, and you don't know whether it's two o'clock or four o'clock or five o'clock in the day because you just keep on going and going and going.

The narratives of people's experiences during the fire demonstrate differences in terms of where people were and what they did. The accounts also offer differential ways in which their bodies, or the bodies of those they were with, coped. For example, one resident recounts how she and her elderly mother had evacuated their home, and, as they were unable to leave Tathra, had sought refuge on the beach. As her mother was fairly immobile, they could not walk further up the beach and had to shelter in a very smoky area:

I was just worried that she was going to have a heart attack, the smoke was so thick… plus, the sand. Just getting sandblasted. So, a lot of the time, we just had to have our backs to the wind with wet T-shirts over our faces. Looking back, all we could see was just thick, black smoke, and spot fires on the beach, on the grass and the shrubs, and just hear gas bottles exploding all over town. So, we just thought everything was gone. Just thought everything down here was burning.

With vision of the town obscured, the previously blue sky turned orange, fires and embers all around and the sounds of explosions, helicopters and planes in the air, people defended and sheltered, terrified that Tathra would be lost to the fire. A number of residents likened the experience to a war zone.

> We were watching the helicopters picking up water right in front of us. We were just watching the plane dropping the retardant stuff. We were hearing helicopters. We were hearing explosions of gas bottles going off. We were just sitting there, just going, "Oh my god. Oh my god. This whole town is going to burn down.

As people reflected on how they had made decisions and navigated during the disaster, their sensory and emotional experiences remained key to understanding what had happened and how they made sense of and navigated the situation. The following resident articulates their experience as one of reacting with instinct and, in the absence of any official information, having to rely solely on themselves. The resident postulates that perhaps self-reliance and making decisions instinctively may sometimes lead to better outcomes. They also describe how their body 'shut-down' giving them tunnel vision and no memory of hearing:

> You're acting on no information whatsoever, so what is it? Your judgment, your instinct, karma...... you just rely on yourself and the decisions that you make. It's like, sometimes it actually is better for people to figure out what they need to do themselves. And that's what happened, because none of us got any warnings... we just all felt very alone... The other thing that I noticed was how in that time of stress, how my body was really shutting down on things that I didn't need. I didn't have peripheral vision... It actually was tunnel vision. It was so strange. And my hearing, I couldn't really hear things. So, it was like really just basic bodily functions to get you through...

Drawn together, these narratives and experiences are illustrative of the intense interaction among senses, emotions and place during the fire. Residents' accounts are illustrative of 'sensuous dispositions' (Paterson, 2009) and of meaning-making from senses of a hazard. These processes arguably shape behaviour and responses during a crisis and are therefore consequential for emergency management. Yet, they are also consequential for thinking through how disorientation is navigated, how various senses and memories are drawn together and combined, and how relationships with/in place are (re)negotiated through sensory engagements.

Quarantined in a ghost town

After the immediate danger of the bushfire had passed, police set up a roadblock to prevent people from entering and returning to Tathra. Still burning fires, downed electricity infrastructure and asbestos posed risks to people's health and safety. Residents who had remained were essentially trapped; if they left, they would not be able to return. Power and phone lines remained off, and communication was limited. Some covertly traversed locally known paths and fire trails into town by vehicle or foot. However, on the whole, for a few days, the residents in Tathra were cut off and isolated from the outside world. People spoke about the eeriness of this post-apocalyptic scene, juxtaposed with comradery and the sharing of food and stories of the fire.

The descriptions of those days in Tathra are evocative, filled with new sounds and sights and the absence of people and electrical power. One resident described the surreal optical changes, the darkness punctuated by the lights from police cars and also the haunting moment when power returned one evening and televisions sprang to life in empty homes. Another resident likens their experience to the famous Australian (book and) film *Tomorrow When the War Began* and that they were the only remaining 'civilians':

> I would describe the whole thing as, what's that movie, trying to think of it now, hang on. Tomorrow When the War Began? It felt that eerie, that it felt like we were the only car in town... we were driving around and then there was just water bombings and just I don't know fire people and yeah, just emergency response people all around the town, like I felt like it was just us. But, I don't know, it's still surreal.

Another resident described a similar experience:

> ... it was just a ghost town. There was a few other people that stayed around town which I knew about, and if we needed a cup... some food, there was no food, no water and everything, so... rifling through people's cupboards and trying to find something to eat and that was pretty tough for the last three days. And even if you wanted to go out to try and get something, they wouldn't let you back in... I didn't have any service on my phone and there was no power to charge it. I had no contact with outside world. That was exactly it, it's like we were in a different world.

A bus tour of Tathra was organised by the local government for those who had evacuated and did not know the status of the damage to their homes. However, they were not permitted to get off the bus as the roadblock was still in place. Residents who had remained in Tathra described how the surreal visit of the bus that was not allowed to stop made them feel as if they were in a 'leper colony':

> This bus came through and I thought oh, they'll stop and talk to us... Then I realized, they're not allowed to stop, they're not even allowed to open the windows... it made me feel a bit like we were in leper colony, look but don't touch.

People talked about how those days felt like a lifetime, a period of transition from their lives before the fire to their new lives post-fire:

> I guess... datewise, it's a couple of days, but it felt like years. It felt like all our lives had changed so much, that when I'm thinking now, I'm thinking, "Oh, that was just Monday, or that was just Tuesday," it's like, "No, it just felt like forever".

Sense of home – sense of safety

Residents discussed the new realisation of the bushfire risk in Tathra, that it was not just people next to the bushland interface directly, or those in more vulnerable homes. The fire's impact and the risk were seemingly random to people, with some vulnerable homes close to the bush having survived and others that seemed outwardly resilient having burnt down. This apparent randomness made the future and their sense of safety and sense of home in Tathra uncertain and unsettling. As this resident says: 'I can't believe how random the fire was, and to see all those wooden houses there, and I'm thinking, "How would they escape when cement-rendered and brick houses were flattened?"' The Bega River that runs between Tathra and the bushland and forest to the north had been thought of as a barrier to bushfire by some, but now people know the fire can easily jump the river and all of Tathra is potentially at risk.

For others, new building codes and asset protection zones mean that their homes, gardens and surrounding landscape will be very different. For many, the changes to the landscape to make communities safer were felt to be 'destroying' the environment that was the essence of Tathra itself:

> That's why I left today, because the guys came and were chopping down trees. I'm like, "Why? It's already been damaged so much." Then a bulldozer came and knocked down even more trees and destroyed the whole hillside... they're like, "Oh yeah, we're fixing things." I'm like, "No, you're destroying things." Destroying is not fixing. It's not the same thing.

Senses of emergency

An implicit goal of emergency management, supported by research into risk perceptions and communication, is to reduce the need for people to rely on their senses (Dash and Gladwin, 2007; Kuligowski and Dootson, 2018). Emergency services aim to provide timely and authoritative warnings and advice to encourage people to take immediate protective action, rather than waiting to be able to sense and confirm the threat themselves. However, personal confirmation (or verification) of the threat through a sensory encounter is a means to reduce decision-making uncertainty. Put simply, the senses assist people to assess the likelihood that they will be impacted and decide if the effort of evacuation or some other protective action (such as defending property) is required (Whittaker et al., 2020). Our findings highlight the important role that senses play in decisions about how to respond to hazards, especially when emergency services' communications are impeded. The importance of people's visual verification of the fire is a recurring theme, however, as discussed by Paterson (2009), the optic senses provide only distal not proximal knowledges, and it is the interaction with the other senses that provides our experience of place. To borrow Fuller's (1732) aphorism (cited in Paterson, 2009), 'seeing is believing... but feeling's the truth'. Others have similarly argued that touch provides a form of confirmation or 'truth' as it generates an awareness of movements, temperatures and pressures that often exceeds the rational and visual and provides us with a distinctive connection with the environment (Hetherington, 2003; Obrador-Pons, 2007). In this way, it can be seen, as Wylie (2005) describes, as an osmotic and soluble becoming of the body and the bushfire elements that then defines a raw human response of flight, flight or shelter.

Many people had not previously considered that they were at risk from bushfire. Tathra was seen as a safe, idyllic environment, where people went for leisure and to retire. In a matter of hours, this idyllic safe haven was transformed into what some people termed 'a war zone'. The familiar and certain became unfamiliar and uncertain. People had to navigate through a changed hazardscape with simultaneously heightened and reduced sensory abilities. For some of those interviewed, Tathra is changed forever. Useful here is the idea of Hetherington (2003: 1939) around 'praesentia' in terms of how place is constructed sensually; how the 'material other', in this case the smoke, heat, wind and embers, can change the feel of a place to one that is unfamiliar and eerily different, providing us with a new 'sense of who we are and where we find ourselves'. The rhetoric of recovery, resilience and building back better, commonly seen in emergency management and governmental responses following disasters, generally all see the bushfire as a temporary presence. However, our interviews show that this is not the case and the fire is ever-present, haunting through the damaged and changed relationships with the more-than-human landscape. Homes and the community are no longer seen as 'safe' and 'idyllic', but at risk; the fire not something restrained by the bush, but able to cross into seaside suburbia. The boundary of what is safe and unsafe is now seen as fluid. Those who loved Tathra for its proximity to the bush now have to come to terms with strict building codes, clear spaces around their homes and more of a tamed, manicured space than the wilderness they once loved.

The sensory engagements of the environment by adults are thought to be numbed (Leder, 1990: in Bartos 2013). However, this work has demonstrated that during a disaster event, adult senses become heightened. As explained by Edensor (2000), quoting Sennett (1994: 309–310), 'the body comes to life when coping with difficulty' and 'is roused by the resistance to which it experiences' (Edensor, 2000: 102). This work will bring more attention to the relationship between embodied sensations and people's movements in a disaster but also importantly to how they feel afterwards, and how it has affected their 'sense of place' and their relationship with their home and community. People's sense of Tathra has changed since the bushfire, and this supports current geographical understanding of 'place' continually 'becoming', with people's relationships to it temporary and changing. This work will also challenge and inform current norms in emergency management, where recovery is thought of as a linear process and once 'rebuilt' the 'place' and people's relationship to it will be unchanged (Okada et al., 2018).

Note

1 This research was made possible through the support of the New South Wales Rural Fire Service, and we particularly thank the staff and volunteers who participated in the interviewing. We also thank the Bushfire and Natural Hazard CRC who assisted through research coordination. We gratefully acknowledge the generosity of the residents of Tathra and Reedy Swamp who shared their stories with us.

References

Anzai S and Kazama SO (2018) Social media analysis of people's high-risk responses to flood occurrence. *Urban Water Systems & Floods* 2: 167–175.

Bartos AE (2013) Children sensing place. *Emotion, Space and Society* 9: 89–98.

Bissell D and Gorman-Murray A (2019) Disoriented geographies: undoing relations, encountering limits. *Transactions of the Institute of British Geographers* 44: 707–720.

Bourdieu P (1987) *Distinction: A Social Critique of the Judgement of Taste.* Cambridge, MA: Harvard University Press.

Classen C (1997) Foundations for an anthropology of the senses. *International Social Science Journal* 49(153): 401–412.

Classen C (1998) *The Color of Angels: Cosmology, Gender and the Aesthetic Imagination.* London: Routledge.

Dash N and Gladwin H (2007) Evacuation decision making and behavioral responses: individual and household. *Nature Hazards Review* 8: 69–77.

Davidson J and Milligan C (2004) Embodying emotion sensing space: introducing emotional geographies. *Social & Cultural Geography* 5: 523–532.

Edensor TIM (2000) Walking in the British countryside: reflexivity, embodied practices and ways to escape. *Body & Society* 6: 81–106.

Fuller, T (1732). *Gnomologia: adagies and proverbs, wise sentences and witty sayings.* London: Barker, and Bettesworth and Hitch.

Haynes, K., Bird, D.K., Carson, D., Larkin, S. and Mason, M. 2011. Institutional response and Indigenous experiences of Cyclone Tracy. Report to National Climate Change Adaptation Research Facility, Gold Coast, Australia. June 2011.

Haynes K, Coates L, Leigh R, et al. (2009) 'Shelter-in-place' vs. evacuation in flash floods. *Environmental Hazards* 8: 291–303.

Haynes K, Tofa M, Avci A, et al. (2018) Motivations and experiences of sheltering in place during floods: implications for policy and practice. *International Journal of Disaster Risk Reduction* 31: 781–788.

Hetherington K (2003) Spatial textures: place, touch, and praesentia. *Environment and Planning A* 35: 1933–1944.

Houston JB, Hawthorne J, Perreault MF, et al. (2015) Social media and disasters: a functional framework for social media use in disaster planning, response, and research. *Disasters* 39: 1–22.

Howes D (1991) *The Varieties of Sensory Experience: A Sourcebook in the Anthropology of the Senses.* Toronto: University of Toronto Press.

Howes D (2019) Multisensory anthropology. *Annual Review of Anthropology* 48: 17–28.

Howes D and Classen C (2013) *Ways of Sensing: Understanding the Senses in Society.* London, New York: Routledge.

Kuligowski E and Dootson P (2018) Emergency notification: warnings and alerts. In: Manzello S (ed.), *Encyclopedia of Wildfires and Wildland-Urban Interface (WUI) Fires.* Cham: Springer.

Leder D (1990) *The Absent Body.* Chicago and London: The University of Chicago Press.

Lund K (2005) Seeing in motion and the touching eye: walking over Scotland's mountains. *Etnofoor* 181: 27–42.

Macpherson H (2009) The intercorporeal emergence of landscape: negotiating sight, blindness, and ideas of landscape in the British countryside. *Environment and Planning A* 41: 1042–1054.

McLuhan M (1962) *The Gutenburg Galaxy.* Toronto: University of 'Ibronto Press.

Obrador-Pons P (2007) A haptic geography of the beach: naked bodies, vision and touch. *Social & Cultural Geography*, 8: 123–141.

Okada T, Howitt R, Haynes K, et al. (2018) Recovering local sociality: learnings from post-disaster community-scale recoveries. *International Journal of Disaster Risk Reduction*, 31: 1030–1042.

Paterson M (2009) Haptic Geographies: ethnography, haptic knowledges and sensuous dispositions. *Progress in Human Geography*, 33: 766–788.

Pred A (1984) Place as historically contingent process: structuration and the time-geography of becoming places. *Annals of the Association of American Geographers* 74: 279–297.

Rodaway P (1994) *Sensuos Geographies: Body, Sense and Place.* London and New York: Routledge.

Sennett R (1994) *Flesh and Stone.* London: Faber.

Stoever JL (2016) *The Sonic Color Line: Race and the Cultural Politics of Listening.* New York: New York University Press.

Straughan E, Bissell D and Gorman-Murray A (2020) The politics of stuckness: waiting lives in mobile worlds. *Environment and Planning C: Politics and Space.* doi: 10.1177/2399654419900189

Tofa M, Haynes K, Avci A, et al. (2017) *Exploring the Experiences of those Who Shelter in Place during Severe Flooding.* Melbourne, VIC: Bushfire and Natural Hazards Co-operative Research Centre.

Whittaker J, Blanchi R, Haynes K, et al. (2017) Experiences of sheltering during the Black Saturday bushfires: implications for policy and research. *International Journal of Disaster Risk Reduction* 23: 119–127.

Whittaker J, Haynes K, Handmer J, et al. (2013) Community safety during the 2009 Australian 'Black Saturday' bushfires: an analysis of household preparedness and response *International Journal of Wildland Fire* 22: 841–849.

Whittaker J, Haynes K, Tofa M, et al. (2018) Reedy Swamp bushfire research. *Report for the New South Wales Rural Fire Service.*

Whittaker J, Taylor M and Bearman C (2020) Why don't bushfire warnings work as intended? Responses to official warnings during bushfires in New South Wales, Australia. *International Journal of Disaster Risk Reduction* 45: 101476.

Wylie J (2005) A single day's walking: narrating self and landscape on the South West Coast path. *Transactions of the Institute of British Geographers* 30: 234–247.

Yusoff K (2007) Antarctic exposure: archives of the feeling body. *Cultural Geographies* 14: 211–233.

13 *Bangla* bricks

Constellations of monsoonal mobilities

Beth Cullen

This chapter traces the meteorological mobilities, entwined within the life course of Bangladesh bricks, also referred to as *Bangla* bricks, to explore how monsoons are embedded within built environments. I follow the object of the brick in order to engage with the agencies that co-produce urban materialities. I focus, particularly, on the role that weather plays in the production and transformation of materials. Mobilities thinking provides a way of exploring and acknowledging how weather, as a form of nonhuman agency, is entangled within social worlds; after all, meteorological forces themselves are both products of and drivers of mobility. Arising from the 'relational movements' (Jones 2011) of the earth, weather patterns shape places, bodies and materials through their movements. As the anthropologist Tim Ingold (2007: 33) points out, people weather their lives by moving alongside atmospheric patterns, and, in the process, 'bind the weather into substantial living forms'. With this in mind, Ingold argues that weather should be thought of as a field of moving materiality which continually gives rise to things and contributes to their dissolution. As meteorological mobilities are entwined with the materials that constitute our environment, he encourages us to 'follow the forces and flows of material' that bring forms into being (Ingold 2010: 10). In doing so, we might gain a deeper appreciation of how materials are caught up in currents of the weather-world.

Following the *Bangla* brick reveals interrelations between mobilities and materialities (Jensen 2016) and the ways in which human and nonhuman forms of mobility are interlinked (Baldwin et al. 2019). This is inspired by a 'more-than-human' approach (Whatmore 2006) that is attentive to the co-constitutive role of nonhumans in the production of the world. Taking inspiration from Ingold, I pay attention to the continual generation and transformation of materials that constitute the brick through 'processes of admixture and distillation, of coagulation and dispersal, and of evaporation and precipitation' (Ingold 2007: 7). The work of Tim Cresswell (2010) provides a way of exploring these processes as he prompts us to consider the forces that constitute movement, notions of speed and rhythm as components of mobility, how mobility is channelled in time and space and the sensory aspects of mobility. By perceiving the brick as a 'constellation of mobility' (Cresswell 2010), connected with flows of varying kinds, I reveal that the lived environment of Bangladesh is 'inextricably bound up with the biophysical forces that transform energy and materials in an active process of co-production' (Taylor 2014: 37). This perspective provides a way of troubling perceived dichotomies between society and meteorological forces.

After a brief overview of the interconnections between weather and brickmaking, and an introduction to the role of bricks as a key component of Bangladesh's built environment, the sections in the chapter follow the life cycle of a *Bangla* brick from its material origins to its dissolution. The section 'Sediment to clay' then explores the intersecting meteorological, geological and human mobilities that produce the clay soils, which constitute the main raw material for brickmaking. 'Clay to brick' focuses on the dry season production process, engaging primarily with the atmospheric entanglements entailed in brickmaking and the human and nonhuman mobilities that these set into motion. Finally, 'Brick to sediment' examines the mobilisation of the brick as a component of road infrastructures within dynamic hydrological landscapes, and the mobilities that these infrastructures enable and constrain. Each section focuses on the complex interactions between humans and nonhumans, mobility and materiality, and the ways in which these interrelations co-constitute social worlds through weathering.

Weather and bricks

From its earliest beginnings, brickmaking has been bound up with the weather. Bricks were formed in relation to weather, created to provide protection from prevalent climatic conditions and insulation from cold and heat (Fernandes et al. 2010). Bricks also respond to weather. Their ability to attract and hold water enables them to absorb moisture from the air and release moisture into the air (Minke 2006), making bricks adaptive to changing climatic conditions. By regulating the microclimates of human habitation, bricks mediate relations with weather and climate. As well as being reactive to weather, bricks are entangled with weather through their production. Historically, brickmaking was an inherently seasonal activity due to its reliance on the heat of the sun. Bricks were made by hand, sun-dried and fired in clamps using firewood and coal. As the entire process was carried out in the open air, it was dependent on the weather but also subject to its uncertainties (Watt 1990: 29). Development of new technologies in Europe and America towards the end of the 19th century transformed the sector (Heierli and Maithel 2008: 29). Brickmaking became large scale, capital-intensive and highly mechanised, diminishing its seasonality.

Relations between weather and brickmaking are still apparent in the monsoonal terrain of Bangladesh, and much of South Asia. In these regions, brickfields only operate during the drier months, stopping just before the rains arrive, the season of production being determined by the monsoon. In its simplest definition, the monsoon is a 'seasonally prevailing wind', the name originating from the Arabic *mausim* meaning season (Fein and Stephens 1987). However, this simple definition belies the monsoon's complexity. Periodic monsoonal winds are created by atmospheric circulations, produced by solar heating and seasonally contrasting temperatures between land and ocean. The immense mountain ranges of the Himalayan-Tibetan Plateau grow warm during spring air rises, drawing in flows of moist air from the oceans to the south. As the air rises, heavy clouds of water vapour form which cool and condense over land as they travel, producing monsoon rains. As the sun retreats and winter approaches, oceans remain warmer than land and the wind direction reverses, carrying rains towards the Indian Ocean. Arising from dynamic interactions between atmosphere, oceans and continents (Clemens et al. 1991), including the spinning earth, transfers of heat between land and sea and the seasonal movement of the sun, the monsoon is a mobile phenomenon generated by 'planetary mobilities' (Szerszynski 2016).

These large-scale monsoonal movements manifest differently in different places. Bangladesh is predominantly influenced by the Bay of Bengal branch of the south-west monsoon. These winds surge across the Bay, arriving in the south-eastern part of Bangladesh by late May or early June. They then move north, crossing the Meghalaya Plateau before arriving at the Himalayas. On reaching the Himalayan foothills, they are deflected west where they advance over the Indo-Gangetic plains. Onset occurs rapidly, with the monsoon covering the entire country from the extreme south-east to the north-west in just 13 days; withdrawal is more variable and proceeds more slowly, beginning in September with the monsoon leaving the country completely by the middle of October (Ahmed and Karmakar 1992). On average, the monsoon lasts for 122 days in Bangladesh (Warrick et al. 2012: 66), although current trends suggest that the rainy season is becoming shorter and the intensity of rainfall is increasing (Ministry of Foreign Affairs of the Netherlands 2018: 4). Despite being broadly predictable, monsoon rhythms fluctuate from year to year and from place to place; they are not metronomic as each season varies from the last, with observations suggesting that monsoon patterns may change in the near future. Nevertheless, the materiality of monsoon winds is undeniable – their presence manifesting in atmospheric humidity, clouds, rain and annual floods (Figure 13.1); their absence marked by heat, dust and dryness.

Oscillations between wet and dry seasons, caused by the periodic movement of monsoon winds, profoundly influence Bangladesh's lived environment. Nearly 80% of the country's annual average rainfall of 1,854 mm falls during monsoon months, with little rainfall the rest of the year (Hossain and Bahauddin 2013). These variations in rainfall create enormous hydrological fluctuations. From November to March, only a fraction of the land is covered in water; however, from June to September, waters swell to cover almost half of the country. This results in a hybrid, shifting terrain where land and water are in constant flux (Lahiri-Dutt 2014). Human and nonhuman

Figure 13.1 Monsoon clouds over Bangladesh.
Source: Beth Cullen.

activities respond to these fluctuations, adjusting to the recurrent transformations of the environment. During the wet season, the monsoon seeps into almost every aspect of life through its smell, wetness, pressure, humidity, puddles, pools and flows (Bremner, forthcoming), becoming entangled with social processes, materialities and practices, including the making of bricks. Exploring the life cycle of the brick within this monsoonal landscape reveals the mutual worlding that Neimanis and Walker (2014) refer to as 'weathering', a process of becoming in which bodies, materials, places and the weather are inter-implicated.

Bangla bricks

The observations outlined in this chapter arise from ethnographic fieldwork in Bangladesh as part of an ongoing research project exploring relations between changing monsoon climates and rapid urbanisation in South Asia. Fieldwork was carried out between February 2018 and October 2019 and consisted of three visits which took place in the dry season, monsoon season and post-monsoon season. During these visits, I spent time observing brickfields, brick markets and the utilisation of bricks in the built environment, mostly on the outskirts of the capital city, Dhaka. I also spoke with brickfield workers and owners, brick suppliers and distributers, architects and planners, researchers and environmental activists.

Brickfields are immediately apparent on visiting Dhaka; they proliferate along the river banks surrounding the city, their towering chimneys emerging from the low-lying riverine terrain. Piles of bricks are transported in overladen boats along waterways and are carried in precarious mountainous stacks by day labourers onto the many construction sites that punctuate the city. Bricks are used for buildings and infrastructure alike. Even Louis Kahn's iconic National Assembly Building (1962–1983) rises up from a base of red *Bangla* bricks, surrounded by a moat signifying the country's aquatic landscape. The foundations and surfaces of the city's roads and pavements are also made from brick, although they often go unnoticed as they are pounded by trucks, bicycle wheels and feet. During the drier months, brick dust infiltrates the air, coating buildings, plants, surfaces and lungs; during wetter months, the monsoon rains dampen the dust, cleansing surfaces and atmospheres alike. Bricks form a ubiquitous, ever-present part of the urban landscape; their visibility and mobility an indicator of rampant urbanisation (Figure 13.2).

Brickmaking in Bangladesh has grown exponentially in recent times. Today, an estimated 7,000 brick kilns produce 23 billion bricks annually (Department of Environment 2017). The sector, which accounts for approximately 1% of the country's GDP and generates employment for more than a million people (Department of Environment 2017), has grown to supply the booming infrastructure and construction industry. Although brick kilns are prevalent throughout Bangladesh, they have mushroomed around Dhaka, one of the world's largest cities. Clusters of brick kilns have sprung up to feed the demand for buildings and infrastructure. In Greater Dhaka alone there are approximately 1,572 active brick kilns, mostly located along canals and rivers which serve as arteries for transportation (Guttikunda et al. 2013). As the city's population continues to grow, with a projected population of 27

Figure 13.2 Piles of clay-fired *Bangla* bricks.
Source: Beth Cullen.

million by 2030 (UN 2016), so too the country's brick industry is expected to grow by 50% within this period (Guttikunda et al. 2013). The continued growth of the city is contingent on the expansion of the brick industry and, as a building material that is deeply entangled with geology, meteorology and hydrology, this has consequences for humans and nonhumans alike.

Sediment to clay: geological mobilities

If we trace the brick to its material origins, we begin with weather. Brickmaking in Bangladesh is reliant on clay-rich soils extracted from the abundant alluvial floodplains of the delta. Although bricks are made from a combination of sand, silt and clay, the essential element for brickmaking is clay. Clay particles lend viscosity and plasticity to clay, enabling it to be moulded into desirable shapes, including the rectangular shape of the brick. The presence of clay particles also causes concretion when heated, transforming plastic soils into strong, water-resistant vitrified forms. Clays are distributed across the globe and exist in the earth's crust, ocean sediments and atmospheric aerosols (Ito and Wagai 2017), but the formation of these extraordinary fine-grained, earthy particles is almost always the result of weathering. Clays occur under a limited range of conditions and are generally found at the surface of the earth at the rock–atmosphere interface (Velde 2008). Most clay materials form when rocks containing feldspar come into contact with water, feldspar being a group of rock-forming crystalline minerals consisting of aluminium silicates (Merriam-Webster n.d.). When large amounts of water are present, the solids in the rock become unstable and they dissolve. The greater the renewal of water, for example by rain, the greater the dissolution (Velde 2008). These erosive movements alter the feldspar minerals resulting in the formation of clay particles. The weathering of rock through movement and friction brings clay into being. Like the monsoon, these tiny, dynamic particles emerge from and are generated by mobilities.

The very existence of Bangladesh is reliant on erosion and the mobility of alluvial materials (Rogers and Overeem 2017). The land that constitutes the Bengal Delta has been formed through the gradual sedimentation of alluvial soils, deposited by rainwater and flowing rivers and shaped by fluvial and tidal processes. On average, the Ganges, Brahmaputra and Meghna rivers are estimated to carry 1 billion tons of sediments to the Meghna estuary each year (Brammer 2016: 43). Only a tiny percentage of this sediment is generated within Bangladesh, much of it originating from the highly erodible slopes of the upper Himalayas (Khalequzzaman 2019). The loose structure of these relatively young mountains makes them particularly vulnerable to erosion (Micheaux et al. 2018). The monsoon plays a significant role, with frequent landslides and fluxes of large sedimentary particles occurring during the rainy season (Struck et al. 2015). Through their weathering by monsoon rains, Himalayan sedimentary rocks feed the clay-rich soils of Bangladesh's humid floodplains. The soils which come to form

Bangladesh bricks were once part of vast mountain systems, a reminder that materials 'have life histories of their own and may have served time in other structures, living and non-living' (Ingold 2004: 240). As Wilson (2017) writes, through the flux and flow of forces and materials and processes of unbecoming, mountains become bricks, roads and architecture.

To move from the mountains to the delta, clay particles require force and the monsoon acts as this agent of mobility. Monsoonal flows routinely and rhythmically shift millions of tons of water and sedimentary material, rhythm being an important component of mobility (Cresswell 2010). Monsoon rains cause the rate of sedimentation and deposition to vary with the seasons and changing velocities of river flow. Fluctuations in flow between wet and dry seasons for the Ganges–Brahmaputra–Meghna Delta are amongst the highest in the world. Rivers that flow sluggishly for most of the year metamorphose into devastating torrents during the monsoon, their waters thick with the enormous amounts of sediment they carry (Lahiri-Dutt and Samanta 2013). The rivers that form the Bengal Delta are highly mobile, changing course over time and depositing sedimentary particles as they meander. As well as changing course, they also change their size, which influences the spatial distribution of sediments. When swollen with monsoon rains, rivers spread out across their floodplains releasing fine clay particles as the speed of their waters slow, leaving deposits of clay in low-lying areas furthest away from the river (Brammer 1996). Whilst the formation of clay soils is dependent on mobility and friction, periods of settling are required too; stillness, or slowness, is also important in worlds of mobility (Cresswell 2014). Fluctuating intensities of flow, intersected by periods of settling, contribute to the rhythmic formation of Bangladesh's clay soils.

The physical processes that contribute to the making of clay soils are regularly interrupted by human interventions. Farmers tend and work soils, transforming them over time into fine-textured, high-yielding soils that are good for cultivation. As a predominantly agrarian economy, Bangladesh is reliant on its soils and, in recognition of their vital role, Bangladeshis refer to soil as mother or *mati* (Sillitoe et al. 2004). However, rapid urbanisation is contributing to soil loss as the country's clay-rich soils are also used for brickmaking (Biswas et al. 2018). Clay is usually mined and prepared near brickfields (Figure 13.3), but as demand rises, it is increasingly sourced from other locations, usually paddy fields. As farmers sell their clay-rich soil to the brick kilns, vast amounts are transported to urban areas. The extent of soil removal is unknown but it is estimated that the brick sector in Bangladesh consumes an estimated 45 million tons of clay annually (World Bank 2011: 17). Soils are collected from a depth of one to two metres (Huq and Shoaib 2013: 126), these top layers usually being the most nutrient-rich. Although the production of all urban materiality involves the movement of material from one place to another (Edensor 2012), in Bangladesh this is precipitating a crisis. Removal of topsoil reduces crop productivity, threatening food security (Roy 2016). Reduced productivity acts as a further incentive for farmers to sell soil (Biswas et al. 2018). As agricultural productivity decreases, movement of people from rural to urban areas increases, swelling the cities (Rahman et al. 2017). As urban populations increase so does demand for housing and infrastructure, fuelling the construction industry and the demand for raw materials.

Due to the speed of urbanisation, removal of topsoil for brickmaking is occurring at a rate that cannot compete with geological cycles. Typically, soil formation takes place over extended time frames. The time needed to form a soil depends on latitude, and the corresponding weather. In wet tropical areas, like Bangladesh, where high temperatures and abundant moisture result in stronger weathering processes, soil formation is faster than in temperate areas (Velde 2008). Nevertheless, the formation of soils is still a slow process. Despite the vast amounts of sediment deposited by monsoonal flows in the Bengal floodplains, it may take 25–50 years for new alluvium to change into soil depending on the location of the land (Brammer 2016). This is compounded by the fact that although Bangladesh is regularly flooded, 'most floodplain land has received little or no new alluvium for several hundred years or more' (Brammer 1996: 26). As a result, a non-renewable resource, on human timescales, is becoming exhausted. Soil has become a tradeable resource, subject to uncontrolled extraction, bought, sold and transported on an unprecedented scale and speed. The channelling of soils into urban landscapes creates frictions between rural and urban, human and nonhuman, and, by facilitating the expansion of the city, the sale of soil generates yet more demand for bricks.

Figure 13.3 Soil preparation for brickmaking.
Source: Beth Cullen.

Clay to brick: atmospheric mobilities

Brickmaking operates around a monsoonal cycle, beginning in November and ending in April before the first monsoon rains arrive (Biswas et al. 2018). The monsoon creates a temporal structure as brickmaking is timed to coincide with the driest part of the year. Brick kilns cannot operate during the wet season as frequent rain, high atmospheric humidity and reduced sunlight affects the drying and firing of bricks (Gomes and Hossain 2003). Water levels rise submerging the brickfields, which are usually situated on low-lying land which is cheaper to buy or to rent because of seasonal flooding. Brickfields become deserted as they transform into pools of water, the slumbering columns of chimneys reflected in their surfaces (Figure 13.4). Although production comes to an end during the wet season, monsoon rains are still leveraged to economic advantage. The price of brick increases during monsoon months as brickfields splutter to a stop, reducing supply and stalling construction. Kiln owners and wholesalers stockpile bricks in the dry season when prices are low, in order to sell in the rainy season when prices rise. Stockpiled bricks fired during the dry season are transported to urban centres during the wet season. Waterways swell with the monsoon, their increased depth enabling the smooth passage of boats. Swollen waterways allow boats to be loaded to capacity making transportation more efficient and profitable. The seasonal fluctuation of riverbanks also alters the distances that bricks have to be carried; the high water levels of the wet season reducing the wages of labourers who carry the bricks from kilns to boats and boats to shore.

During the monsoon, boats are also hired to carry soil and coal, the essential raw materials for brickmaking, at much lower costs than land

Figure 13.4 Submerged brick kiln during wet season.
Source: Beth Cullen.

transportation. These materials are mounded and stored on areas of high land adjacent to the kilns ready for the dry season. Clay soils are collected and weathered during the wet season in preparation for the dry season. Open-air stockpiles sit throughout the monsoon months, exposed to the action of the atmosphere. The heaps are regularly broken up, cut and turned over so that the atmosphere may penetrate them. During this period of stasis, rainwater spreads through the body of clay, diffusing and softening it, increasing its workability and making it pliable for moulding. The action of wind, rain and air help to homogenise the clay, breaking down harder lumps and agglomerates and washing out unwanted soluble salts. Weathering is used to work the clay, reducing human labour. Once the rains draw to a close, the water levels slowly start to recede, exposing the brickfields and allowing the brickmaking cycle to begin again.

Brickfields churn relentlessly in the drier months, transforming landscapes and atmospheres. During this period, bricks are made by hand, sundried, fired and then cooled. Clay is a heavy material that requires a lot of energy to work. Each brick is created through gruelling, repetitive labour, formed by the respirations, pulses and circulations of human bodies. Sand is added to temper the clay before being moulded to increase its plasticity and workability. Tempered clay is then processed, or pugged, to ensure its homogeneity, which prevents bricks from cracking when drying. Once the clay is prepared, moulding can begin. Balls of pugged, tempered clay are thrown forcibly into wooden moulds before being skilfully turned out onto sanded, levelled ground (Figure 13.5). Moulding is repeated through rhythmic replications producing endless rows of unfired, green bricks which are left to dry in the sun until they are firm enough to be fired, the drying dependent on circulations of the atmosphere. In the labour-intensive process of brickmaking, it is not just the clay that is transformed, people change too, 'their bodies moulded to the daily tasks, their senses attuned to the subtle "voices" of the machines and matter they are working with' (Bennett 2016: 72). Brickmaking is a co-creation of the affordances of matter and human energies, and in the process, people are weathered too.

Firing is the final stage of the brickmaking process. Heat imparts strength and hardness to the brick, converting clay from a soft ephemeral material into a durable, weather-resistant form. Brickmaking is a kind of alchemy, 'the elements of earth and water are transformed by fire into a material that can be more durable than stone' (Cruickshank 2019: no page). Kiln technologies lie at the heart of the firing process, and in Bangladesh, most bricks are produced using Fixed-Chimney Bull's Trench Kilns (Department of Environment 2017), which are designed for continuous production. Batches of bricks are fired and cooled in seemingly endless repetitions. Kilns are tended by workers who feed them with fuel at regular intervals. The structures have no permanent roof, which is why they cannot operate during the monsoon, meaning that kiln top workers are exposed to intense heat from the flames below and from the sun above. During firing temperatures reach

Figure 13.5 Brick moulding and sun-dried bricks.
Source: Beth Cullen.

700–1,100°C, intensifying the heat of the dry season, subjecting workers to dehydration, heat stroke and skin and eye diseases (Maithel et al. 2014). Vitrification occurs between 900 and 1150 °C (Campbell and Pryce 2003: 14), causing clay particles to fuse together. In doing so, firing changes the temporal structure of the clay, producing a material that can outlive the precarious lives of those who bring it into being.

Bricks depend on human labour and the brickfields of Bangladesh on seasonal migrants. Labour migration is partly structured by the monsoon cycle. Men, women and children make their way to Dhaka's brick kilns from all over the country. Migrants arrive at the start of the dry season, when there is limited agricultural work, and return to their home villages at the start of the monsoon season when the brickfields stop operating. Workers migrate from the poorest areas of Bangladesh, which are affected in various ways by monsoonal forces. Some come from so-called *monga* areas which suffer from shortages of rainfall, some from the northern plains and char islands which are subject to flash floods and riverbank erosion, and some from the cyclone- and storm surge–vulnerable south. Migration is a response to a dynamic monsoonal environment that requires mobility for survival (Ingham et al. 2019). The migrant labourers who power the brickfields of Bangladesh are a reminder that not everyone weathers equally (Neimanis and Hamilton 2017). As poverty, ecological degradation and changing weather patterns intensify, there has been a dramatic increase in the spatial mobility of Bangladesh's population. A large proportion migrate to Dhaka, the already bursting capital city, which is estimated to receive 300,000–400,000 people every year (World Bank 2007: xiii). Urban expansion offers abundant opportunities for labourers who carry sand, bricks and cement; break bricks; and produce building materials for the ongoing production of the city.

Although bricks are essential building materials, and a source of income for many, their production contaminates urban atmospheres (Figure 13.6). Dhaka has some of the most polluted air in the world, regularly topping global rankings of cities with the worst air quality (Dhaka Tribune 2019). Particulates from the brickfields have been found to constitute an estimated 30–40% of ambient PM2.5 pollution over the Dhaka Metropolitan Area (Guttikunda et al. 2013). Pollution levels are exacerbated by inefficient kiln technologies (Department of Environment 2017). Plumes of black smoke, blankets of thick smog and clouds of brick dust emitted by the kilns flux with the seasons; the movement of particulates influenced by the movement of the weather. Air quality monitoring has revealed that concentrations peak in December and January, partly due to the slow-moving winds that occur during these months reducing dispersion (Guttikunda et al. 2013). The tiny airborne particles penetrate the lungs and the bloodstream, causing cardiovascular and respiratory conditions (Khan 2019). High concentrations of particulate pollution cause an estimated 2,200–4,000 premature deaths per year in the greater Dhaka region (Saha and Hosain 2016: 491). Pollutants affect nonhumans too; dry brick kiln dust dispersed by the wind reduces photosynthesis and hinders plant growth, damages soils and erodes building surfaces (Darain et al. 2013). The very bricks that build the city make it unliveable for humans and nonhumans alike.

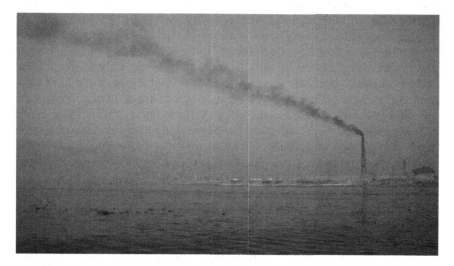

Figure 13.6 Emissions from brick kilns during dry season.
Source: Beth Cullen.

Pollutants from the brickfields do not remain confined to urban environments, they also affect atmospheres on a regional scale. The primary source of fuel for brick kilns is coal, annual consumption of which exceeds 1 million tonnes (Department of Environment 2017). Through their emission of CO_2, black carbon and tropospheric ozone, brick kilns affect atmospheric temperatures, monsoon circulation and rainfall patterns (Mitra and Valette 2017). Such particulates contribute to the Asian brown cloud, a mobile atmospheric phenomenon consisting of sulphates, nitrates, organics, black carbon, fly ash and other pollutants, that occupies the air space over much of South Asia every year (Begum et al. 2011; Sharma et al. 2019). The brown cloud appears between early December and March, coinciding with the dry season and the brickmaking period, although there are many other contributing factors. This mobile assemblage of pollutants mingles with atmospheric circulations, becoming entangled with meteorological processes. While the monsoon modifies the mobility and severity of the haze by washing pollution particles out of the atmosphere during the rainy season, it seems the atmospheric brown cloud in turn modifies the monsoon, changing the energy balance of the atmosphere which influences rainfall distribution (Liepert and Giannini 2015). The mobile particles released by the brickfields contribute to changing weather patterns and the mobilities of the human and nonhuman lives that are enmeshed within them.

Brick to sediment: hydrological mobilities

After firing, *Bangla* bricks are graded, their classification determined by the evenness of firing, their colour, strength and the presence of cracks or breaks. Depending on their classification, they are allocated for use in a diverse range of buildings and infrastructure. Around 15–20% of bricks are overfired or exposed to excessive heat, which can cause defects, so they are broken into small pieces called pickets (Luby et al. 2015). It is incredible to observe the energies that go into the making of bricks only to watch them being broken up, but these pieces of brick are a vital part of the built environment. Pickets are used extensively in road construction and form a base for nearly all paved roads in the country (Luby et al. 2015). Thick layers of brick chips are dispersed along the burgeoning highway network. These layers are flattened and compacted by steam rollers before being covered in tarry layers of bitumen. Bricks are used because boulders, gravel and stone are in short supply. As rivers move further from their source, the size of the material they carry gets smaller, with larger boulders, rocks and stones getting ground down as they are transported. This means that although Bangladesh has an abundance of fine-grained clay, silt and sand, it has few stones. Gravel is only available from a few quarries, mostly located in northern Sylhet (Rahman 2009). The distant location of the quarries leads to high haulage costs, so bricks are used instead. Daily labourers, often migrants, are employed to manually crush bricks, striking them repeatedly with a metal hammer until desired sizes are obtained. Mimicking the forces of weathering and erosion, brick-breakers produce an alternative to the stones that Bangladesh is deprived of by monsoonal flows.

Roads are an indicator of economic development and changing modes of mobility. Historically, the main transport routes in Bangladesh were waterways (Quium and Hoque 2002). As in other places, roads have largely replaced water transportation, partly due to the rise of motorised vehicles and the improvement of overland surfaces (Paterson 2014). Construction of 'all-weather roads' is now a development priority for Bangladesh as they allow movement in all seasons. The Asian Development Bank recently allocated 200 million dollars to improve the rural road network in Bangladesh as only 40% of the rural population currently has access to all-weather roads (Daily Star 2019). Such roads are constructed not to be flooded, sodden or muddy and are partly a response to the sticky, clayey mud that accumulates in the monsoon season making movement difficult. It is somewhat ironic that these roads are formed from the very material that acts to constrain mobility. Once constructed, their presence creates new geographies and lines of relation, facilitating the movement of people, goods and materials. National development plans have placed an emphasis on the construction of highways which are intended to improve connectivity (ADB 2014), particularly with the southern coastal districts, paving the way for development and ongoing expansion into the delta.

In the construction of these roads, bricks that have been subject to intensities of heating are used to elevate and protect them against intensities of wetness. Strategies of topographic alteration are particularly important in Bangladesh as two-thirds of the land area is no higher than five metres above sea level (Adams et al. 2011). During the monsoon, water levels swell to inundate almost half of the country, hindering terrestrial mobility. As a result, roads are typically constructed to ensure that their surfaces lie above wet season flood levels. In this mostly flat delta terrain, even small changes in elevation make a big difference. 'All national and regional roads in Bangladesh were designed to be built above the highest flood level... and feeder roads were designed above normal flood level' (Dasgupta et al. 2010: 12). Roads are elevated by layering materials, this typically includes a compacted soil foundation, a sub-base, a road base and a surface layer (Rahman 2009), with bricks comprising an important component of these arrangements (Figure 13.7).

Although roads create new speeds and rhythms of movement, they can also contribute to unexpected and unplanned flows and resistances (Merriman 2016). The majority of roads in Bangladesh do not have adequate drainage facilities for flood water (Dasgupta et al. 2010), meaning flows of water and sediment are obstructed. As a result, roads constructed to facilitate human mobilities can hinder and disrupt nonhuman mobilities (Fishel 2019). There is growing recognition that linear infrastructures, which have become pervasive features of anthropogenic landscapes, can affect ecological functions

Figure 13.7 Road construction using *Bangla* bricks.
Source: Beth Cullen.

and processes, including the movement of water across landscapes (Raiter et al. 2018). Impacts can be subtle but may extend over large areas, well beyond the direct infrastructure footprint (Raiter et al. 2018). An example from Bangladesh is the Dhaka–Rajshahi Highway which was constructed through Chalan Beel, one of the country's largest wetlands. Although the highway reduced the distance between the capital and Rajshahi by 75 kilometres (Ali 2008), it also dissected the wetland, altering its hydrology. The highway is the most recent in a series of embanked infrastructures, including the colonial-era Eastern Bengal Railway Line, the construction of which was also dependent on bricks (John 2018). Over time, these structures have significantly obstructed water flows, hastening the silting up of the *beel* (a shallow, seasonal wetland), reducing its water-holding capacity and significantly impacting regional ecologies (Iqbal 2019).

As brick-built structures, roads in Bangladesh are prone to processes of 'molecular mobility' (Merriman 2016), slowly eroding over time through a series of often barely perceptible micro-events. Interactions between water and earth are the primary catalyst for their decline. As such, most roads in Bangladesh experience regular deterioration during the rainy season. Seasonal fluctuations of wetness cause the underlying clay soils to shrink and swell, weakening road foundations. Weathering by monsoon rains creates cracks which allow water to seep inside, small cracks become large potholes which are enlarged by the pummelling of traffic and the action of wind and rain. Water absorption is one of the main causes of road deterioration and the porosity of the brick chips that comprise the primary aggregate material leads to high rates of absorption (Mazumder et al. 2006). Once inside, water

Figure 13.8 Weathered road surface and disintegrated bricks.
Source: Beth Cullen.

acts a lubricant, breaking down particle bonds, a process that is exacerbated when the road is subjected to dynamic human and nonhuman moving forces. As a 'thing of the earth', the brick is 'forever undergoing a process of decay, of crumbling back into earth' in a slow yet relentless process of decomposition (Steele and Vizel 2013: 82). Through weathering and disintegration, brick returns to sediment (Figure 13.8), creating seasonal cycles of maintenance and repair and fuelling the never-ending demand for brick.

Conclusion

Following the materiality of the brick reveals the multiple ways in which the monsoon is enmeshed within lived environments in Bangladesh. As more-than-human assemblages, bricks come into being not through human agency alone, but through entanglements of earth, sky, wind, weather and human energies. Weathering contributes to the emergence of bricks, their production, use and dissolution, and the interconnected human and nonhuman mobilities caught up in these processes. As my study of *Bangla* bricks has revealed, far from being a disconnected backdrop to social life, meteorological forces are co-constitutive of social worlds, moving through them and the materials that compose them (Neimanis and Walker 2014). Through its generative movements, the mobile materiality of the monsoon is entwined within the very building blocks of Bangladesh's cities and the infrastructures on which they depend.

Focusing on the meteorological mobilities embedded within the life course of Bangladesh bricks, hopefully goes some way towards challenging perceived dichotomies between society and meteorological forces. Learning to consider weather patterns, like the monsoon, as vital actors in our entangled mobile world is an urgent necessity in light of increasingly uncertain futures. As Barry (2018) points out, grappling with the diversity of materials in motion is vital for life in the Anthropocene. It is essential that we seek to better understand our immersion within materially complex mobilities and our co-constitutive role in these arrangements. The 'geosocial formations' (Clark and Yusoff 2017), of which we are part, propagate through the environment, their trajectories unpredictable. Caught up in the flows and fluxes of the weather-world, even something as seemingly static and bounded as the brick is a vibrant, mobile entity capable of changing the geologies, atmospheres and hydrologies we inhabit.

References

Adams, N, Dasgupta S and Sarraf, M (2011). *The Cost of Adapting to Extreme Weather Events in a Changing Climate*. Bangladesh Development Series, Paper No. 28. Washington, DC: World Bank. http://documents.worldbank.org/curated/en/716231468014361142/The-cost-of-adapting-to-extreme-weather-events-in-a-changing-climate

ADB (2014). Bangladesh: Road Maintenance and Improvement Project. *Performance Evaluation Report*. www.adb.org/documents/bangladesh-road-maintenance-improvement-project

Ahmed, R and Karmakar, S (1992). Arrival and Withdrawal Dates of the Summer Monsoon in Bangladesh. *International Journal of Climatology*, 13: 727–740.

Ali, A (2008). Toll Paying Commuters Suffer as Natore-Sirajganj Road Rundown in Five Years. *The Daily Star*, June 21. www.thedailystar.net/news-detail-42172

Baldwin, A, Frohlich, C and Rothe, D (2019). From Climate Migration to Anthropocene Mobilities: Shifting the Debate. *Mobilities*, 14(3): 289–297.

Barry, K (2018). More-than-human Entanglements of Walking on a Pedestrian Bridge. *Geoforum*, 106: 370–377.

Begum, B, Biswas, S, Pandit, G, Saradhi, I, Waheed, S, Siddique, N, Seneviratne, M, Cohen, D, Markwitz A and Hopke, P (2011). Long-range Transport of Soil Dust and Smoke Pollution in the South Asian Region. *Atmospheric Pollution Research*, 2(2): 151–157.

Bennett, L (2016). Thinking like a Brick: Posthumanism and Building Materials. In: Taylor, C and Hughes, C (eds.), *Posthuman Research Practices in Education*. Palgrave Macmillan: London, 58–74.

Biswas, D, Gurley, E, Rutherford, S and Luby, S (2018). The Drivers and Impacts of Selling Soil for Brick Making in Bangladesh. *Environmental Management*, 62: 792–802.

Brammer, H (1996). *The Geography of the Soils of Bangladesh*. Dhaka: University Press Limited.

Brammer, H (2016). *Bangladesh: Landscapes, Soil Fertility and Climate Change*. Dhaka: University Press Limited.

Bremner, L (Forthcoming). Monsoon Assemblages. In: Ashraf, K (ed.), *Locations: Anthology of Architecture and Urbanism Volume 2*. Dhaka: Bengal Institute for Architecture, Landscapes and Settlements.

Campbell, JWP and Pryce, W (2003). *Brick: A World History*. London: Thames and Hudson.

Clark, N and Yusoff, K (2017). Geosocial Formations and the Anthropocene. *Theory, Culture & Society*, 34(2–3): 3–23.

Clemens, S, Prell, W and Murray, D (1991). Forcing Mechanisms of the Indian Ocean Monsoon. *Nature*, 353: 720–725.

Cresswell, T (2010). Towards a Politics of Mobility. *Environment and Planning D: Society and Space*, 28(1): 17–31.

Cresswell, T (2014). Friction. In Adey, P, Bissell, D, Hannam, K, Merriman, P and Sheller, M (eds.), *The Routledge Handbook of Mobilities*. London: Routledge, 107–115.

Cruickshank, D (2019). The First Cities. In: Hall, W (ed.), *Brick*. London: Phaidon Press Limited.

Daily Star (2019). ADB provides $200m to upgrade rural road network in Bangladesh. *Daily Star*, January 13. www.thedailystar.net/banking/rural-road-development-in-bangladesh-adb-provides-200-million-dollars-upgrade-1686970

Darain, K, Rahman, A, Ahsan, A, Islam, A and Yusuf, B (2013). Brick Manufacturing Practice in Bangladesh: A Review of Energy Efficiency and Air Pollution Scenarios. *Journal of Hydrology and Environment* Research, 1(1): 60–69.

Dasgupta, S, Huq, M, Khan, ZH, Masud, Md, Ahmed, M, Mukherjee, N and Pandey, K (2010). *Climate Proofing Infrastructure in Bangladesh. The Incremental Cost of Limiting Future Inland Monsoon Flood Damage*. Policy Research Working Paper 5469. Washington, DC: World Bank. http://documents.worldbank.org/curated/en/387911467997269676/Climate-proofing-infrastructure-in-Bangladesh-the-incremental-cost-of-limiting-future-inland-monsoon-flood-damage

Department of Environment (2017). *National Strategy for Sustainable Brick Production in Bangladesh*. Department of Environment, Government of the People's Republic of Bangladesh. www.ccacoalition.org/en/resources/national-strategy-sustainable-brick-production-bangladesh

Dhaka Tribune (2019). Air Quality Index Ranks Dhaka's Pollution 3rd Worst in World. *Dhaka Tribune*, May 9. www.dhakatribune.com/bangladesh/dhaka/2019/05/09/air-quality-index-ranks-dhaka-s-pollution-3rd-worst-in-world

Edensor, T (2012). Vital Urban Materiality and its Multiple Absences: The Building Stone of Central Manchester. *Cultural Geographies*, 20(4): 447–465.

Fein, JS and Stephens, PL (eds.) (1987). *Monsoons*. New York: John Wiley.

Fernandes, FM, Lourenco, PB and Castro, F (2010). Ancient Clay Bricks: Manufacture and Properties. In: Dan, M, Prikryl, R and Torok, A (eds.), *Materials, Technologies and Practice in Historic Heritage Structures*. Dordrecht: Springer, 29–48.

Fishel, SR (2019). Of Other Movements: Nonhuman Mobility in the Anthropocene. *Mobilities*, 14(3): 351–362.

Gomes, E and Hossain, I (2003). Transition from Traditional Brick Manufacturing to More Sustainable Practices. *Energy for Sustainable Development*, 7(2): 66–76.

Guttikunda, SK, Begum, BA and Wadud, Z (2013). Particulate Pollution from Brick Kiln Clusters in the Greater Dhaka Region, Bangladesh. *Air Quality Atmosphere and Health*, 6: 357–365.

Heierli, S and Maithel, U (2008). *Brick by Brick: The Herculean Task of Cleaning up the Asian Brick Industry*. Berne: SDC.

Hossain, N and Bahauddin, K (2013). Integrated Water Resource Management for Mega City: A Case Study of Dhaka City, Bangladesh. *Journal of Water and Land Development*, 19: 39–45.

Huq, I and Shoaib, J (2013). *The Soils of Bangladesh*. World Soils Book Series. Dordrecht: Springer.

Ingham, V, Islam, MR and Hicks, J (2019). Adaptive Flood Mobilities in Bangladesh. *Mobilities*, 14(2): 158–172.

Ingold, T (2004). Buildings. In: Harrison, S, Pile, S and Thrift, N (eds.), *Patterned Ground: Entanglements of Nature and Culture*. London: Reaktion Books, 238–240.

Ingold, T (2007). Earth, Sky, Wind and Weather. *The Journal of the Royal Anthropological Institute*, 13(1): S19–S38.

Ingold, T (2010). *Bringing Things to Life: Creative Entanglements in a World of Materials*. NCRM Working Paper Series. ESRC National Centre for Research Methods. http://eprints.ncrm.ac.uk/1306/1/0510_creative_entanglements.pdf

Iqbal, I (2019). Railways in Colonial Bengal. *The Daily Star*, April 8. www.thedailystar. net/in-focus/news/railways-colonial-bengal-1726765

Ito, A and Wagai, R (2017). Global distribution of Clay-size Minerals on Land Surface for Biogeochemical and Climatological studies. *Scientific Data*, 22: 4.

Jensen, O (2016). Of 'Other' Materialities: Why (Mobilities) Design Is Central to the Future of Mobilities Research. *Mobilities*, 11(4): 587–597.

John, J (2018). *Archaic Technology, Social Relations and Innovations in Brick Kilns*. New Delhi: Centre for Education and Communication.

Jones, J (2011). Lunar-solar Rhythmpatterns: Towards the Material Culture of Tides. *Environment and Planning E: Economy and Space*, 43(10): 2285–2303.

Khalequzzaman, Md (2019). Transboundary River Flow: The Future of Bangladesh Depends On It. *The Daily Star*, July 18. www.thedailystar.net/opinion/news/the-future-bangladesh-depends-it-1772926

Khan, M (2019). The Dust-laden Air of Dhaka. *The Daily Star*, April 5. www.thedailystar. net/star-weekend/environment/news/the-dust-laden-air-dhaka-1724887

Lahiri-Dutt, K (2014). Beyond the Water-Land Binary in Geography: Water/Lands of Bengal Re-Visioning Hybridity. *ACME: An International Journal for Critical Geographies*, 13(3): 505–529.

Lahiri-Dutt, K and Samanta, G (2013). *Dancing with the River: People and Life in the Chars of South Asia*. New Haven: Yale University Press.

Liepert, B and Giannini, A (2015). Global Warming, the Atmospheric Brown Cloud, and the Changing Indian Summer Monsoon. *Bulletin of the Atomic Scientists*, 71(4): 23–30.

Luby, SP, Biswas, D, Gurley, E and Hossain, I (2015). Why Highly Polluting Methods are used to Manufacture Bricks in Bangladesh. *Energy for Sustainable Development*, 28: 68–74.

Maithel, S, Kumar, S and Lalchandani, D (2014). *Factsheet about Brick Kilns in South and South-East Asia*. New Delhi: Greentech Knowledge Solutions Pvt Ltd.

Mazumder, A, Kabit, A and Yazdani, N (2006). Performance of Overburnt Distorted Bricks as Aggregates in Pavement Works. *Journal of Materials in Civil Engineering*, 18(6): 777–785.

Merriam-Webster Online Dictionary (n.d.). Feldspar. accessed January 7, 2020. www.merriam-webster.com/dictionary/feldspar

Merriman, P (2016). Mobilities II: Cruising. *Progress in Human Geography*, 40(4): 555–564.

Micheaux, F, Mukherjee, J and Kull, C (2018). When Hydrosociality Encounters Sediments: Transformed Lives and Livelihoods in the Lower Basin of the Ganges River. *Environment and Planning E: Nature and Space*, 1(4): 641–663.

Ministry of Foreign Affairs of the Netherlands. 2018. Climate Change Profile Bangladesh. The Hague: Ministry of Foreign Affairs. www.government.nl/binaries/government/documents/publications/2019/02/05/climate-change-profiles/Bangladesh.pdf

Minke, G (2006). *Building with Earth: Design and Technology of a Sustainable Architecture*. Berlin: Birkauser.

Mitra, D and Valette, D (2017). *Brick by Brick: Unveiling the Full Picture of South Asia's Brick Kiln Industry and Building the Blocks for Change*. Geneva: International Labour Organisation, Brooke Hospital for Animals and the Donkey Sanctuary. Neimanis and Hamilton.

Neimanis, A and Walker, RL (2014). Weathering: Climate Change and the "Thick Time" of Transcorporeality. *Hypatia*, 29(3): 558–575.

Paterson, M (2014). Carbon. In: Adey, P, Bissell, D, Hannam, K, Merriman P and Sheller, M (eds.), *The Routledge Handbook of Mobilities*. London: Routledge, 326–334.

Quium, A and Hoque, A (2002). The Completeness and Vulnerability of Road Network in Bangladesh. In: Ali, A, Seraj, S and Ahmed, S (eds.), *Engineering Concerns of Flood: A 1998 Perspective*. Dhaka: Bangladesh University of Engineering and Technology, 59–75.

Rahman, A (2009). Use of Brick Aggregate for Road Constructions in Bangladesh. MSc. Thesis. Bangladesh University of Engineering and Technology.

Rahman, M, Hassan, S, Bahauddin, K, Ratul, A and Bhuiyan, M (2017). Exploring the Impact of Rural-Urban Migration on Urban Land Use and Land Cover: A Case of Dhaka City, Bangladesh. *Migration and Development*, 7(2): 222–239.

Raiter, K, Prober, S, Possingham, H, Westcott, F and Hobbs, R (2018). Linear Infrastructure Impacts on Landscape Hydrology. *Journal of Environmental Management*, 206: 446–457.

Rogers, K and Overeem, I (2017). Doomed to Drown? Sediment Dynamics in the Human-Controlled Floodplains of the Active Bengal Delta. *Elementa: Science of the Anthropocene*, 5: 66.

Roy, S (2016). Precious Topsoil Burned for Bricks. *The Daily Star*, January 26. www.thedailystar.net/frontpage/brick-kilns-breed-woes-farmers-207520

Saha, CK and Hosain, J (2016). Impact of Brick Kilning Industry in Peri-Urban Bangladesh. *International Journal of Environmental Studies*, 73(4): 491–501.

Sharma, S, Khanna, I and Ghosh, P (2019). *Scoping Study for South Asia Air Pollution*. New Delhi: The Energy and Resources Institute (TERI). www.gov.uk/dfid-research-outputs/scoping-study-for-south-asia-air-pollution#citation

Sillitoe, P, Barr, J and Alam, M (2004). Sandy-Clay or Clayey-Sand: Mapping Indigenous and Scientific Soil Knowledge on the Bangladesh Floodplains. In: Bicker, A, Sillitoe, P and Pottier, J (eds.), *Development and Local Knowledge*. London: Routledge, 174–200.

Steele, WE and Vizel, I (2013). Housing and the Material Imagination – Earth, Fire, Air and Water. *Housing, Theory and Society*, 31(1): 76–90.

Struck, M, Andermann, C, Hovius, N, Korup, O, Turowski, JM, Bista, R, Pandit, HP and Dahal, RK (2015). Monsoonal Hillslope Processes Determine Grain Size-Specific Suspended Sediment Fluxes in a Trans-Himalayan River. *Geophysical Research Letters*, 42(7): 2302–2308.

Szerszynski, B (2016). Planetary Mobilities: Movement, Memory and Emergence in the Body of the Earth. *Mobilities*, 11(4): 614–628.

Taylor, M (2014). *The Political Ecology of Climate Change Adaptation: Livelihoods, Agrarian Change and the Conflicts of Development.* London: Routledge.

UN (2016). *The World's Cities in 2016- Data Booklet.* Department of Economic and Social Affairs, Population Division. New York: United Nations. www.un.org/en/development/desa/population/publications/pdf/urbanization/the_worlds_cities_in_2016_data_booklet.pdf

Velde, B (2008). Geology of Clays and Earthen Materials. In: Avrami, E, Guillaud, H and Hardy, M (eds.), *Terra Literature Review: An Overview of Research in Earthen Architecture Conservation.* Los Angeles: The Getty Conservation Institute, 8–14.

Warrick, RA, Azizul Hoq Bhuiya, AK and Mirza, MQ (2012). The Greenhouse Effect and Climate Change. In: Warrick, RA and Ahmad, QK (eds.), *The Implications of Climate and Sea-Level Change for Bangladesh.* Dordrecht: Kluwer Academic Publishers, 35–96.

Watt, K (1990). Nineteenth Century Brickmaking Innovations in Britain: Building and Technological Change. *PhD Thesis.* Institute of Advanced Architectural Studies, University of York.

Whatmore, S (2006). Materialist Returns: Practicing Cultural Geography in and for a More-Than-Human World. *Cultural Geographies,* 13(4): 600–609.

Wilson, K (2017). Untimely Mountains | Entangled Matter. In: Kakalis, C and Goetsch, E (eds.), *Mountains, Mobilities and Movement.* London: Palgrave Macmillan, 171–186.

World Bank (2007). *Dhaka: Improving Living Conditions for the Urban Poor.* Bangladesh Development Series, Paper No. 17. Washington, DC: World Bank. http://documents.worldbank.org/curated/en/938981468013830990/Dhaka-Improving-living-conditions-for-the-urban-poor

World Bank (2011). *Introducing Energy-efficient Clean Technologies in the Brick Sector of Bangladesh.* Report No. 60155-BD. Washington, DC: World Bank. http://siteresources.worldbank.org/BANGLADESHEXTN/Images/295758-1298666789983/7759876-1323201118313/BDBrickFINAL.pdf

14 Weathering colonisation

Aboriginal resistance and survivance in the siting of the capital

Sarah Wright, Lara Daley and Faith Curtis

In this chapter, we aim to learn from weather, its beings, agencies and co-becomings. Our intention is to highlight the contingencies of settler colonialism (Snelgrove et al., 2014), and to contribute to the generative potential of weather and weathering. In settler colonial contexts such as Australia, place has been weathered in deeply racist, entitled and possessive ways across time and space (Watson, 2005; Moreton-Robinson, 2015). Yet, Indigenous peoples, nations and Country itself persist, weathering the damage of ongoing colonisation and continuing to know and live with weather in diverse and place-based ways. Engaging weather in this chapter, we heed the call of Warlpiri man Wanta Steve Jampijinpa Patrick (2015), who invites Aboriginal and non-Aboriginal people today to learn from *Milpirri*, the cloud that forms when hot air and cold air meet and interact, the stormcloud of different knowledges meeting. Jampijinpa Patrick shares that while there is difficulty in these interactions, they can lead to new understandings and with this, rain, which can result in a nourishment of Country.

We, Sarah, Lara and Faith, three non-Indigenous academics, write this chapter from the unceded lands of Awabakal, Worimi and Gumbaynggirr Country. As academics on stolen lands, we are always situated at the interface of different knowledges meeting (Nakata, 2007) and are trying to find ways to respectfully learn from our own interactions with Indigenous knowledges. In our engagements we hope that we might, like the rain of *Milpirri*, be part of the nourishing potential of interactions across difference, acknowledging that these understandings have arisen from colonising histories and presents, in which we are also implicated. In writing this paper, we acknowledge the diverse weather cultures of multiple places and co-becomings that weave through our thinking weather in this chapter and enable our existence in place. We also acknowledge the disruptions of those same weather cultures that are also a part of how we come to be(ing) here (Snelgrove et al., 2014). It is in following Jampijinpa Patrick's place-conscious 'ethical relationality' (Donald, 2012) that we wish to turn up the heat on the way that weather can be enrolled in ongoing colonisation, and respectfully engage weather beings, agencies and co-becomings.

Paying attention to weather and the ways it shapes human movement (Birtchnell and Büscher, 2011; Whatmore, 2013; Vannini, 2016), is differentially experienced (Neimanis and Hamilton, 2018) and co-becomes as place (Bawaka Country including Wright et al., 2019) is a relatively new area for social science research (Vannini et al., 2012). Yet there is nothing new about attending to weather for the many Indigenous lifeworlds that have engaged the beings, co-becomings and agencies of weather for millennia. As cautioned by Metis and Otipemisiwak feminist scholar Zoe Todd (2016: 8), 'new' social science research must take care not to treat climate as a form of *aer nullius* – 'a blank commons to be populated by very Euro-Western theories of resilience, the Anthropocene, Actor Network Theory', among others (see also Juanita Sundberg, 2014, on the more-than-human/post-human). Taking direction from Todd, to ignore or sideline complex Indigenous relationships with climate is to risk re-colonising Indigenous epistemological and ontological domains.

In response, Indigenous scholarship calls for engagements with weather to be contextualised in this historical experience and in place-based political relationships. Indigenous thinkers such as Vanessa Watts (2013) and Zoe Todd (2016) point out that diverse Indigenous cosmologies, 'habitats and ecosystems' (Watts, 2013: 23), as well as 'climate and atmospheres' (Todd, 2016: 6), form part of human and more-than-human social organisation and legal-political orders (Napoleon, 2007). For us, following Todd and many other Indigenous thinkers, the practices of weathering and the beings, co-becomings and agencies of weather are never politically neutral – they emerge in and through place (Graham, 2001; Coulthard, 2010, 2014; Watts, 2013; Larsen and Johnson, 2017). In settler colonial contexts, such as Australia, weathering place is entangled in processes of ongoing colonisation.

In such contexts, weather can become enrolled in political projects of (re)possession, ordering and control of people, place and time(s), yet it can also be an active agent of resistance to this. Therefore, in this chapter, we aim to think of weather in radically contextual (Howitt, 2011) and place-based ways that are more ably positioned to engage with the power geometries of weathering places of Indigenous survivance and ongoing colonisation. In taking an approach that grounds weather in place, we draw attention to the political connotations of weathering and how weathering relates directly to projects of both settler colonialism and Aboriginal and Torres Strait Islander sovereignty in Australia.

We draw attention to settler colonialism by considering the political debates surrounding the process of choosing a site for the capital of Australia at the time of federation (1901). The choice of the capital was seen as a defining moment of nation-building; for creation of the colonial settler state. Indeed, this was to be a central move in the establishment and creation of the newly imagined colonial nation of Australia, part of a raft of policies and laws designed to produce a white, capitalist state (Jayasuriya et al., 2003). In this process of selection and possession, the Aboriginal peoples whose Countries were being discussed and potentially 'selected' including

Ngunnawal, the Walgalu, Ngarigo and the Wuradjeri nations and others, were invisibilised, their land and connections and ways of knowing, being and relating to the world, their very selves and families, violently erased through the pick-and-choose of colonial privilege and drive for possession (Coulthard, 2014; Moreton-Robinson, 2015). In selecting a site with a 'not undesirable' climate, lay an attempt to enrol weather in this invasion, an enactment of immense entitlement and violence.

In their assessment of what was required of a site that would support colonial governance, colonists mobilised ideas around climate that privileged their right to be in place and to govern. In expounding the weather-places that would support colonial governance, colonists actively invisibilised the fact that diverse weather-places were already supporting human and more-than-human systems of governance and legal orders. In many instances, this 'invisibilisation' also involved direct violence in the form of murders, massacres and forced removals, extreme measures taken with the aim of severing relationships between people and place (see for example the map of Ryan et al., 2018 for this devastating history). Weathering place, then, was enrolled in the eliminatory logic of terra nullius, attempting to empty places of their human and more-than-human beings, political agencies and capacities. Yet diverse Aboriginal and Torres Strait Islander governances, supported by place and weathering in different place-based ways were always there and continue to be a part of weather-places.

By focusing on this particular moment in settler nation-building, then, we aim to situate weathering in Australia in the wider and deeper relations of Indigenous survivance and ongoing colonisation, emphasising the way that weather is part of shaping who and what belongs to/with/as nations. Grounding our thinking in the case of selecting the colonial nation's capital, we examine weather's enrolment in possession but also attest to the continuance and resistance of weather beings, co-becomings and agencies that can never be erased.

Weathering place and placing weather

Weather has long been enrolled in the colonial project. While likely imagined as a backdrop to colonisation, accounts by would-be colonisers in many different contexts are redolent with weather's active agencies' and colonisers' (often-thwarted) efforts to enrol or manage weather to support imperial and colonial projects (Naylor, 2006; Mahony, 2016). Janković (2006) points out that the development of climate science and meteorology formed part of this effort, as the emerging science of meteorology aimed to know weather, potentially rendering it measurable, predictable and even controllable. In this context, meteorology worked to develop 'laws' that might support and enable colonial expansion and enterprise (see also Achbari and van Lunteren, 2016; Mahony, 2018).

In considering climate knowledges in Australia, O'Gorman traces the ways that climate science was foundational in establishing norms related to weather in support of colonial expansion; seeking to make weather useful in a wide range of situations from water management, colonial agriculture and government planning (O' Gorman, 2014; see also Beattie, O' Gorman and Henry, 2014; O' Gorman et al., 2016). As O'Gorman et al. point out, 'Colonial weather and climate knowledge had an essential and fundamentally practical purpose – to ensure settlement's success' (2016: 898). Certainly, in the siting of the new capital, a move redolent with both symbolic and perceptible power, weather was a key, and highly contested, actant.

Prior to 1901, Australia was a collection of six British colonies that were self-governing, each under the power of the British Parliament. Australia became an independent nation on 1 January 1901, when the British Parliament passed legislation allowing the six colonies to govern in their own right as part of the Commonwealth of Australia. A key moment in this story of settler colonial nation-building was the decision of where to locate the new federal capital.

The process of choosing a site for the colonial capital was both prolonged and fraught. The two dominant cities of Sydney and Melbourne would not agree to the other being the site of the colonial capital. In 1899, a compromise was reached – included as section 25 of the Constitution – that the capital would be no less than 100 miles (160 kms) from Sydney, while Melbourne would act as the interim capital (Ling, 2013: 7). The words of the Chief Justice of the Commonwealth of Australia framed the issue as follows:

> The Federal Capital should be central, easily accessible, not unduly exposed to the risk of war or invasion, and its climate should not be such as to render it an undesirable place to live.
>
> (Chief Justice of the Commonwealth of Australia, 1986, cited in House of Reps 8 October 1903)

The colonising weathering of places enrolled in selecting a site for government imagined that particular weather-places were more suited to the practices of governing than others. The criteria listed to manage the process served to reduce the complex sovereignties and interplay of weather's beings and becomings into manageable parts; framing the process as technocratic and technoscientific. According to the Report from the Royal Commission and subsequent debates, the criteria used to determine each of the potential sites were as follows:

> Accessibility from various capital cities; Means of communication; Climate; Topography; Water Supply; Drainage; Soil; Building Materials; Fuel; General Suitability of Site; Cost of resumption of site and area and Crown Lands available; General.
>
> (Royal Commission on Sites for the Seat of Government, 1904: iii)

Within the Climate category, the following things were to be considered:

> Elevation above sea-level; Rainfall (amount and how distributed through the year); Temperature (mean for each month, as well as maximum and minimum) and Frequency of fog, mist, snow, and hot or cold winds.
>
> (1904: iii)

In this Royal Commission's category, the diverse agencies of weather are indeed entangled with and as place, as a potential site is understood and grasped through climatic processes. Weather, too, its beings and becomings, the fog, mist, snow, wind, temperatures and rain, are undeniably accorded a powerful agency – the potential to make or unmake the desirability of different sites and so the imagined colonial settler nation itself. Yet, these connections are marshalled in deliberately colonising ways – in order to possess, make and remake a white patriarchal space-as-nation (Moreton-Robinson, 2015). Such a move to possession, as Moreton-Robinson (2015) points out, is predicated on the dispossession of others; dispossession of land, of Country and of the relationships, communications and sovereignties of weather. The notion of desireability (or "not undesirable" as the Chief Justice suggested), is a desire of possession, the desire to create certain desirable subjects and subjectivities. This is a desire for – and indeed a deliberate, violent construction of – a white nation, something that, in the minds of its proponents, would be propelled in its realisation by a white, cold capital. As Senator Pearce conjectured of Bombala (Ngarigo Country):

> There is no objection to the summer climate of Bombala; but it is said of the winter climate that it is subject to bleak cold winds, with occasional falls of snow. This Senate is largely, or at any rate to some extent, composed of men, or the sons of men, who came from the United Kingdom. To which site is there the greater objection? To a site with a bleak winter, or a site with a humid summer? Is it not a fact that the white race reaches its greatest perfection in cold and bleak countries rather than in countries of humid atmospheres.
>
> (Senator Pearce, 14.10.1903)

Australia is being made in these spaces and interactions, through powerfully interwoven tropes of weather, race and colonialism. In doing so, colonists ignored and violently tried to erase extant Aboriginal and Torres Strait Islander systems of governance and legal orders that were already in and as place. To be perfect is to be cold, bleak, to be a son of the United Kingdom. Science and scientists (white, male, cold, bleak scientists, it goes without saying) are once again enrolled in the process. This is clear in the first debates of the Australian Parliament's House of Representatives. As one Member of Parliament (MP) argued:

> If scientists were appealed to they would with one voice attest that those who are born and reared in localities at a considerable elevation, and in a cool climate, are hardier than are those of the same race brought up in the humid climate of the sea coast. A change from one climate to the other is good at all times, and that is one of the reasons why I wish to have a Capital site from which easy access may be had to the sea coast.
> (Mr Clarke MP 08.10.1903)

Indeed, while science, with its unifying predictions (the 'one voice' of those scientists) and its measurable outcomes, is marshalled, there is also a strongly visceral, emotive and embodied experience within the discussions. The weather makes people in particular ways that are bound deeply with imagined histories of self and nation. As Mr O'Malley MP pontificates:

> The history of the world shows that cold climates have produced the greatest geniuses, all of whom were born north of a certain degree … In conclusion, let me say this: look where we like, it will be found that wherever a hot climate prevails, the country is revolutionary. Take the sons of some of the greatest men in the world, and out them into a hot climate like Tumut or Albury, and in three generations their lineal descendants will be degenerate … I want to have a cold climate chosen for the capital of this Commonwealth. I am glad that the Minister for Defence has put that point before us.
> (Mr O'Malley MP 08.10.1903)

Mr O'Malley is not only marshalling knowledges that look to 'the' history of the world, tracing genealogies and deploying 'facts' – but speaking with visceral emotion, evoking a certain colonial white supremacist dream of co-becoming, of cold with perfection, of bleakness with genius, as if the weather can be enrolled to make the subjects of this country – seemingly all white, male subjects – geniuses; perfect, non-degenerate, hardy in their bracing whiteness.

Yet, there is fear, too, and a deep vulnerability. The treasured hardiness of these men is peculiarly fragile, threatened by the sea, by humidity and by heat, by Aboriginal people, by women, by people of colour, by people from continental Europe, by anything other than cold, high-altitude temperate climates peopled by sons of the United Kingdom. There is, then, a fear threaded through the quotes, a fear of becoming differently, a fear of becoming something else. If these settlers were to integrate into this place, what might happen? It seems the underpinning their proclamations, is an awareness that this place might change them, if they cannot change it first. For Mr O'Malley and Mr Clarke, for Senator Pearce and all the rest, the siting of the capital is a point of resistance, an attempt to resist Country changing them by weathering it in certain particular ways. Such anxiety, as Moreton-Robinson (2015) powerfully points out, always lurks with possession, a symptom of attempting to possess that which is not yours.

Possession, then, is never finished, never complete but always must be defended, remade; just as colonialism, the creation of the settler state of Australia, is never complete but always emerging (Moreton-Robinson, 2015; Tuck and Yang, 2012). The capital, chosen eventually in 1908 to be Yass-Canberra, has never actually been settled in any finished and complete way (Simpson, 2014; Howitt, 2019). And as colonialism is never finished, is a process rather than a structure, asserted in continuing ways, so survivance and Country are always there. In the moment of making, there was and always is unmaking. In the desperation and efforts to enrol weather, weather agencies, along with Aboriginal people who were invisibilised but very much there, the women, people of colour – all beings that defied, refused, lived and live outside this fantasy – persisted.

Agencies of weather and Aboriginal weathering

In 1913, Canberra became the official name of the Yass-Canberra area chosen by the Commonwealth Parliament to be the site of the national capital (CBR Canberra, 2017). Yet, this newly termed Canberra was already a place of Aboriginal history, nations and sovereignty. The many weather-places of so-called Australia have always supported human and more-than-human Aboriginal and Torres Strait Islander systems of governance and legal orders. The Country/ies that colonists debated to be the seat of colonial government were also already places of deep political agency, manifested as Aboriginal nations. These places always have and always will continue to nourish and sustain in spite of colonial disruptions (Watts, 2013; Simpson, 2014; Moreton-Robinson, 2015).

Today, Canberra itself is a clear example of the political nature of Indigenous survivance. Despite being chosen as the seat of colonial government with the strong and expressed intention of securing white privilege, it remains a place of ongoing Indigenous political expression and strong refusal. The Aboriginal Tent Embassy Movement, which has seen embassies appear across the continent of what is now Australia, started in Canberra in 1972 with one of the most significant Aboriginal land rights protest actions in Australia of the 20th century (Pieris, 2012; Foley et al., 2014). Iveson (2017) has examined how the Canberra Tent Embassy, situated on crown lands, asserts an enduring sovereignty that the Australian state wishes to deny. Iveson writes that the Embassy's ongoing 'presence remains both an embodiment of Aboriginal sovereignty and a thorn in the side of those whose authority to regulate the space is founded on colonial dispossession' (2017: 3). Still situated in the nation's capital, the Embassy is a long-standing reminder that Aboriginal and Torres Strait Islander legal orders are still here, asserting Aboriginal sovereignty in the present.

Moreover, the Aboriginal Tent Embassy in Canberra continues to inspire the expression of Indigenous sovereignty in and as place across the continent. As we write this chapter, the Djap Wurrung Heritage Protection

Embassy and their allies are standing with 800-year-old sacred trees on Djap Wurrung Country, protecting them from 'a planned highway extension set to destroy this dreaming landscape' (Djap Wurrung Embassy, 2019). A number of the trees they are protecting have been used by generations of women as a place to give birth. The Djap Wurrung Embassy is enacting custodial obligations and responsibilities, protecting sacred trees who have nurtured and supported the re-birthing and co-becoming of their nation across an estimated 50 generations (Martin, 2019). Country holds and acts with these political movements in multiple ways, including with and as the urban Country of Canberra, the ancestor trees of Djap Wurrung Country and the beings, becomings and sovereignties of weather.

For place supports and empowers resistances and survivances – this is something emphasised by many Indigenous thinkers including Glenn Coulthard (2010), Larsen and Johnson (2017), and Daigle (2016) who suggest that, as well as a way of knowing, being and living the world, place may guide resistance and call those who may listen to its defence. As Glenn Coulthard of the Yellowknives Dene First Nation (2010) points out: Indigenous ways of knowing, experiencing place

> ...often guide forms of resistance to power relations that threaten to erase or destroy our senses of place. This, I would argue, is precisely the understanding of land and/or place that not only anchors many Indigenous peoples' critique of colonial relations of force and command, but also our visions of what a truly post-colonial relationship of peaceful co-existence might look like.
>
> (2010: 79–80)

This is also emphasised by Cree geographer Michelle Daigle (2016), who discusses the way specific place-based more-than-human relationships are foundational to practices of self-determination.

Taking direction from Indigenous thinking, more-than-human agencies, and the beings, becomings and sovereignties of weather, empower resistances to colonisation, support the ongoing realisation of ways of being beyond colonialism, demand struggle and provide vision for transformation. Weather acts politically and has the capacity to resist and evade capitalist-colonial enclosures. As Mr Austin Chapman MP concedes, speaking in the House of Representatives on 8 October 1903:

> We should first demand those essentials which money cannot buy, such, for instance, as a good climate ... money can secure a water supply, whereas it cannot purchase a good climate.
>
> (Mr Austin Chapman MP, 8.10.1903)

Money cannot buy the beings and becomings of weather (nor water for that matter, Mr Chapman). In spite of efforts to order and contain, weather is unruly, evading regimes of colonial control. The beings and becomings of weather must be lived-with in order to co-exist.

Aboriginal and Torres Strait Islander ways of knowing and living with the beings, becomings and sovereignties of weather have nourished such co-existence in the diverse weather-places of Australia throughout millennia. The fact that money cannot buy a climate also reveals further agencies of weather in resisting colonisation. In the case of choosing a site for government, for example, the particular agency of heat burns strong. In a senate session on 28 October 1908, Senator Givens argued:

> I recollect that in his first speech which he declared that one-half of Australia was too hot for white men to live in. Apparently he now wishes to argue that the other half is too cold. Are we a tropical race? Are we not accustomed to frost and cold? Are not the residents of Dalgety fine, sturdy, specimens of our race?
>
> (Senator Givens, 28.10.1908)

Senator Givens' comments reflect the opinion of many of the colonists in these debates that much of Australia was 'too hot to live in'. In this way, the heat of Country in many places acted to protect those places from particular colonial presences, in this case from becoming the governing centre of state colonial power. The heat of Country has the capacity to resist colonial presence. As the quote from Senator Givens reflects, heat limits the mobility of particular bodies across particular places. Heat can harm, dehydrate and burn but it can also nourish places, placing limits and restrictions on entry and thus holding space for multiple Aboriginal and Torres Strait Islander nations against an invading one. Mr Conroy MP, speaking in the House of Representatives, unwittingly confesses the fragility of white men in claiming that 'one can protect himself against the cold, but not against heat' (08.10.1903). The heat of Country confounds the hopes of whiteness:

> I want to have a climate where men can hope. We cannot have hope in hot countries. When I go down the streets of this city on a hot summer's day and see the people in a melting condition, I look upon them with sorrow and wish I were away in healthy Tasmania. I hope that the site selected will be Bombala, and that the children of our children will see an Australian Federal city that will rival London in population, Paris in beauty, Athens in culture, and Chicago in enterprise.
>
> (Mr O'Malley MP, 08.10.1903)

In this quote from Mr O'Malley, heat sears the hopes of the colonial project. The resistances and agencies of Country, such as heat, thwarted those early colonial hopes and continue to do so today. Heat finds its way to support survivances and resistances in places and ways early colonists were unable to conceive of; heat in its intensity of weather that melts the hopes of whiteness, heat of fires that co-nourish self-determination and resistances.

Consequently, weather supports everyday survivances, as Indigenous practices of resurgence and self-determination are supported by, guided by, the beings and becomings of weather and place. In and as place, through everyday life, through the messages of particular winds and weathers, Indigenous practices of resurgence are guided and nourished; and in ways deeply connected to place, to sovereignty, to Indigenous belongings and legal orders that have, literally, come into being with and as Country. The charcoal of fires burnt and burning through millennia, for example, found in the layers of land itself in all Aboriginal and Torres Strait Islander nations of Australia, is a more-than-human manifestation of Indigenous sovereignty and co-becoming, of land-as-self-as-Country (Bawaka Country including Wright et al., 2016).

Believed to be a place that would support colonial governance, Canberra continues to be Country that supports Aboriginal nations with heat, and other agencies of weather, nourishing everyday self-determination and igniting resistance in different ways. And, while these relationships and agencies of weather are place-based, they also travel through kinship networks, through the mobile agencies of weather and heat. The sacred fire at the Tent Embassy, for example, burns as part of the political capacities of Country in ways that have spread across Aboriginal nations. This is both a spread of ideas and knowledges, something pointed out by Anoma Peiris in her discussion of the conceptual mobility of the Tent Embassy (2012), as well as a connecting onwards through kinship to other places and communities, and, further, a more-than-human mobility as agencies themselves travel. Through the Tent Embassy movement, the sacred fire travels, carried across nations, igniting place-based resistance with other Country/ies, interacting with other resistances and agencies, human and more-than-human (see for example Daigle, 2016; Daley, 2019). Thus heat, fire and place all nourish resistance in different places, in different ways, refusing to be extinguished by colonial weatherings of place.

The burning of the sacred fire at embassies across many nations and place-based struggles in Australia also invites Indigenous and non-Indigenous people to gather, forge new relationships and resist together. As Daley (2019: 303) has noted from Aboriginal activists' reflections in Meanjin (Brisbane), the sacred fire – as part of ceremony, song, dance and yarning – gathers and situates 'the coming into relation of non-Indigenous people with Indigenous struggles on Indigenous terms, awakening people's hearts and souls and generating new pathways of struggle'.

The responsibilities to heed weather in different ways are not responsibilities of Indigenous people alone (Daigle and Sundberg, 2017). Weather may also create conditions for dialogue and transformation for Indigenous people *and* settler/migrant-colonisers, nourishing more productive ontological relationships with place. As Bawaka Country et al. (2016) suggest, speaking of winds, seasons and *lirrwi* [the charcoal]:

> *Ŋapaki* [non-Indigenous people] may see the *lirrwi* [charcoal], may challenge themselves to realize its significance, to listen to the calls of Country, attend to the wind and the new season and the messages they bring, to assemble and recognise their more-than-human kin. For to recognize the existence of diverse ontologies is not a matter of a superficial multiculturalism but requires a re-understanding of the self and the world, a different way of relating, not only to difference, but to the inner-most essence of what is known and how and why ... To do so will lead to radically different ways of being and knowing, and radically different politics, emerging differently in different place-based ways, through the smoke of the fire, creating new levels of *lirrwi*.
>
> (2016: 26)

Weather, then, calls those who would heed it into different kinds of more-than-human responsibilities and struggle albeit, in different ways as we all can only ever start from, speak from, learn from and co-become from and with our own positionally, our own place (Larsen and Johnson, 2017; Smith et al., 2018). In doing so, place may also lead to different kinds of decolonising subjectivities. For non-Indigenous people, this may mean being open to the ways place may change us, being open to the very processes that so terrified Mr Chapman and colleagues.

As the stormcloud

How weather is known, is deployed, manipulated, enrolled was and, indeed, is central to the colonial project, the settler state. Who belonged/belongs; how they feel; how they think and act; what this nation is, was and will be are part of the colonial project. In the case of selecting a colonial capital, elites tried to enrol the weather in this way, to privilege themselves, to make themselves and to realise the colonial project.

We have focused in this chapter on an historical event, the siting of Canberra, the capital of a new colonial settler state. We have done so not to locate colonising weathering in the past, quite the opposite. We feel it is important to both acknowledge these overt moments in ongoing dispossession and to track continuities, particularly to amplify the ways they continue to inform, underpin and empower certain understandings and responses to climate change. As O'Gorman et al. (2016: 898) observe, the production of scientific climate knowledges has long been deployed to promote colonisation in settler colonial contexts, with weather knowledges adding to a suite of colonising ways of knowing, categorising and controlling environments (O'Gorman, 2014: 181). These ways of knowing continue to be enmeshed in today's weather knowledges as data and institutions built during early colonisation continue to 'shape the nature and use of climate knowledge' (O'Gorman et al., 2016: 894). In these ways, settler colonialism continues to shape and be shaped by climate knowledge – what it is and what is known, what is measured, and how it is responded to as indicated in colonial data to inform present forecasting and research. Settler colonialism also shapes and is shaped in present institutions, not least the federal government, that continues to enact the colonial project, through trajectories that began with invasion, dreams of cold whiteness, and that continue to be supported through vast and complex epistemological, ontological frames, deeply imbued with colonial/capitalist/exclusionary/violent logics.

Yet, weather is much more than this. As Indigenous scholar Vanessa Watts (2013) guides us, speaking from her own Anishnaabe and Haudenosaunee cosmologies, weather is a part of so-called natural systems, which from an Indigenous point of view are

> Better understood as societies…meaning that they have ethical structures, interspecies treaties and agreements, and further their ability to interpret, understand and implement. Non-human beings are active members of society. Not only are the active, they also directly influence how humans organize themselves into that society.
>
> (2013: 23)

Following Watts, the diverse beings, co-becomings and agencies of weather such as rain, thunder, heat, air, snow and frost can each be understood as political agencies within complex spiritual, legal and political relations. Consequently, bringing Indigenous epistemologies and ontologies to bear on understandings of weather, contextualises weather in place and in relationships, revealing its deeply political nature.

For us, Sarah, Lara and Faith, learning to learn our responsibilities, the sacred fires burning across nations and *Milpirri*, the stormcloud that is formed when different knowledges come together, act as resistances and survivances but also invitations to form new political relationships on the terms of Aboriginal people and Country; invitations that demand responding and a willingness to sit around the fire or enter the stormcloud to work to engender different ways of being, knowing and doing at the interface of Indigenous and non-Indigenous domains (Nakata, 2007).

In Australia, Aboriginal and Torres Strait Islander peoples and their more-than-human relatives have been weathering intense changes and disruptions to their life-worlds and legal orders for hundreds of years. Growing global knowledges of climate change makes visible what can be better understood as an intensification of ongoing colonial violence (Hatfield et al., 2018). For Indigenous peoples surviving ongoing genocide, climate change is not the first existential crisis to be faced. Nor is it the first life-world-altering change of climate to be lived through. Aboriginal people have already weathered the end of an ice age and sea-level rises on this continent (Nunn and Reid, 2016). There is incredible strength and ancestral knowledge in people and Country, running deep in time and place. It is through a respectful engagement with these knowledges and the restoring of Aboriginal Law/lore to its proper and rightful place as First Law on these lands, that the colonial violence underpinning the destruction of multiple life-worlds might genuinely be addressed.

References

Achbari, A and van Lunteren, F (2016). 'Dutch Skies, Global Laws', *Historical Studies in the Natural Sciences,* 46, 1–43.

Bawaka Country, including Wright, S, Suchet-Pearson, S, Lloyd, K, Burarrwanga, L, Ganambarr, R, Ganambarr-Stubbs, M, Ganambarr, B and Maymuru, D (2016). 'The Politics of Ontology and Ontological Politics', *Dialogues in Human Geography,* 6(1), 23–27. doi: 10.1177/2043820615624053.

Bawaka Country, including Wright, S, Suchet-Pearson, S, Lloyd, K, Burarrwanga, L, Ganambarr, R, Ganambarr-Stubbs, M, Ganambarr, B and Maymuru, D (2019). 'Gathering of the Clouds: Attending to Indigenous Understandings of Time and Climate through Songspirals', *Geoforum,* Online access through: doi: 10.1016/j.geoforum.2019.05.017

Beattie, J, O' Gorman, E and Henry, M (2014). 'Climate, Science, and Colonization: Histories from Australia and New Zealand', In Beattie, J, O' Gorman, E and Henry, M (Eds.), *Climate, Science, and Colonization: Histories from Australia and New Zealand* (pp. 1–16). New York: Palgrave Macmillan.

Birtchnell, T and Büscher, M (2011). 'Stranded: An Eruption of Disruption', *Mobilities,* 6(1), 1–9.

Coulthard, G (2010). 'Place against Empire: Understanding Indigenous Anti-Colonialism', *Affinities: A Journal of Radical Theory, Culture and Action,* 4(2), 79–83.

Coulthard, G S (2014). *Red Skin, White Masks: Rejecting the Colonial Politics of Recognition.* Minneapolis: University of Minnesota Press.

Daigle, M (2016). 'Awawanenitakik: The Spatial Politics of Recognition and Relational Geographies of Indigenous Self-determination', *The Canadian Geographer,* 60(2), 259–269.

Daigle, M and Sundberg, J (2017). 'From Where We Stand: Unsettling Geographical Knowledges in the Classroom', *Transactions of the Institute of British Geographers,* 42(3), 338–341.

Daley, L (2019). *An Urban Cultural Interface: (Re)thinking Urban Anti-capitalist Politics and the City in Relation to Indigenous Struggles.* University of Newcastle, Australia. Unpublished PhD thesis, Access: lara.daley@newcastle.edu.au

Djap Wurrung Embassy (2019). *Protect Sacred Trees.* Available at: https://dwembassy.com/.

Donald, D (2012). 'Indigenous Métissage: A Decolonizing Research Sensibility', *International Journal of Qualitative Studies in Education,* 25(5), 533–555. doi: 10.1080/09518398.2011.554449.

Donald, D (2012). 'Indigenous Métissage: A Decolonizing Research Sensibility', *International Journal of Qualitative Studies in Education,* 25(5), 533–555.

Foley, G, Schaap, A and Howell, E (2014). 'Introduction', In Foley, G, Schaap, A and Howell, E (Eds.), *The Aboriginal Tent Embassy: Sovereignty, Black Power, Land Rights and the State* (pp. xxv–xxxi). Abingdon, Oxon: Routledge.

Graham, M (2001). 'Understanding Human Agency in Terms of Place: A Proposed Aboriginal Research Methodology', *PAN: Philosophy Activism Nature*, 6, 71–78.

Hatfield, SC, Marino, E, Whyte, KP, Dello, KD and Mote, PW (2018). 'Indian Time: Time, Seasonality, and Culture in Traditional Ecological Knowledge of Climate Change', *Ecological Processes*, 7(25), 1–11. doi: https://doi.org/10.1186/s13717-018-0136-6

Howitt, R (2019). 'Unsettling the Taken (-for-granted)' *Progress in Human Geography*, 44(2), 193–215. doi: 10.1177/0309132518823962.

Howitt, R (2011). 'Knowing/Doing', In Del Casino, VJ et al. (Eds.), *A Companion to Social Geography* (pp. 131–145). Chichester, West Sussex; Malden, MA: Wiley-Blackwell.

Iveson, K (2017). "Making Space Public' Through Occupation: The Aboriginal Tent Embassy, Canberra', *Environment and Planning A*, 49(3), 537–554. doi: 10.1177/0308518X16682496.

Jampijinpa Patrick, WS (2015). 'Pulya-ranyi Winds of Change', *Cultural Studies Review*, 21(1), 121–31.

Janković, V (2006). 'The End of Classical Meteorology, c. 1800', *Geological Society Special Publication*, 256, 91–99.

Jayasuriya, L, Walker, D and Gothard, J (2003). *Legacies of White Australia: Race, Culture and Nation*. Crawley, WA: University of Western Australia Press, 2003. Crawley, WA: University of Western Australia Press.

Larsen, SC and Johnson, JT (2017). *Being Together in Place: Indigenous Coexistence in a More Than Human World*. Minneapolis: University of Minnesota Press.

Ling, T (2013). *Government Records about the Australian Capital Territory*. Canberra: National Archives of Australia. http://guides.naa.gov.au/records-about-act/index.aspx

Mahony, M (2016). 'For an Empire of 'all types of climate': Meteorology as an Imperial Science', *Journal of Historical Geography*, 51, 29–39.

Mahony, M (2018). 'The 'genie of the storm': Cyclonic Reasoning and the Spaces of Weather Observation in the Southern Indian Ocean, 1851–1925', *British Journal for the History of Science*, 51, 607–633.

Martin, L (2019). 'Protesters Defend Sacred 800-Year-Old Djap Wurrung Trees as Police Deadline Looms', *The Guardian (Australian Edition)*, 22 August. Available at: https://www.theguardian.com/australia-news/2019/aug/22/protesters-defend-sacred-800-year-old-djap-wurrung-trees-as-police-deadline-looms.

Moreton-Robinson, A (2015). *The White Possessive: Property, Power and Indigenous Sovereignty*, Minneapolis: University of Minnesota Press.

Nakata, M (2007). *Disciplining the Savages, Savaging the Disciplines*, Canberra: Aboriginal Studies Press.

Napoleon, V (2007). *Thinking about Indigenous Legal Orders*. National Center for First Nations Governance.

Naylor, S (2006). 'Nationalizing Provincial Weather: Meteorology in Nineteenth-Century Cornwall', *The British Journal for the History of Science*, 39(3), 407–433.

Neimanis, A and Hamilton, JM (2018). 'Weathering', *Feminist Review*, 118, 80–84.

Nunn, PD and Reid, NJ (2016). 'Aboriginal Memories of Inundation of the Australian Coast Dating from More than 7000 Years Ago', *Australian Geographer*, 47(1), 11–47.

O' Gorman, E (2014). '"Soothsaying" or "Science ?": H. C. Russell, Meteorology, and Environmental Knowledge of Rivers in Colonial Australia', In Beattie, J, O' Gorman, E and Henry, M (Eds.), *Climate, Science, and Colonization: Histories from Australia and New Zealand* (pp. 177–193). New York: Palgrave Macmillan.

O' Gorman, E, Beattie, J and Henry, M (2016). 'Histories of Climate, Science, and Colonization in Australia and New Zealand, 1800–1945', *WIREs Climate Change*, 7(November/December), 893–909. doi: 10.1002/wcc.426.

Pieris, A (2012). 'Occupying the Centre: Indigenous Presence in the Australian Capital City', *Postcolonial Studies*, 15(2), 221–248.

Royal Commission on Sites for the Seat of Government (1904). *Report of the Commissioners*. Sydney: William Applegate Gullick, Government Printer.

Ryan, L, Richards, J, Pascoe, W, Debenham, J, Anders, R, Brown, M, Smith, R, Price, D, Newley, J (2018). *Colonial Frontier Massacres in Eastern Australia 1788. 1930, v2.1* Newcastle: University of Newcastle, 2018, https://c21ch.newcastle.edu.au/colonialmassacres (accessed 3/10/2019).

Simpson, A (2014). *Mohawk Interruptus: Political Life across the Borders of Settler States*. Durham, NC: Duke University Press.

Smith, AS, Smith, N, Wright, S, Hodge, P and Daley, L (2020). 'Yandaarra is living protocol', *Social & Cultural Geography*, 21(7), 940–961. doi: 10.1080/14649365.2018.1508740.

Snelgrove, C, Dhamoon, RK and Corntassel, J (2014). 'Unsettling Settler Colonialism: The Discourse and Politics of Settlers, and Solidarity with Indigenous Nations', *Decolonization: Indigeneity, Education & Society*, 3(2), 1–32.

Sundberg, J (2014). 'Decolonizing Posthumanist Geographies', *Cultural Geographies*, 21(1), 33–47.

Tuck, E and Yang, KW (2012). 'Decolonization Is Not a Metaphor', *Decolonization: Indigeneity, Education & Society*, 1(1), 1–40.

Todd, Z (2016). 'An Indigenous Feminist's Take on the Ontological Turn: "Ontology" Is Just Another Word for Colonialism', *Journal of Historical Sociology*, 29(1), 4–22.

Vannini, P (2016). 'Storm Watching: Making Sense of Clayoquot Sound Winter Mobilities', In T Duncan, SA Cohen and M Thulemark (Eds.), *Lifestyle mobilities: Intersections of Travel, Leisure and Migration* (pp. 209–222). Oxon; New York: Routledge.

Vannini, P, Waskul, D, Gottschalk, S and Ellis-Newstead, T (2012). 'Making Sense of the Weather: Dwelling and Weathering on Canada's Rain Coast', *Space and Culture*, 15(4), 361–380.

Watson, I (2005). 'Settled and Unsettled Spaces: Are We Free to Roam?' *Australian Critical Race and Whiteness Studies Association Journal*, 1, 40–52.

Watts, V (2013). 'Indigenous Place-Thought & Agency amongst Humans and Non-humans (First Woman and Sky Woman Go On a European World Tour!)', *Decolonization: Indigeneity, Education & Society*, 2(1), 20–34.

Whatmore, SJ (2013). 'Earthly Powers and Affective Environments: An Ontological Politics of Flood Risk', *Theory, Culture and Society*, 30(7/8), 33–50.

15 Dwelling and weather

Farming in a mobilised climate

Gail Adams-Hutcheson

Introduction

Climate is fundamentally fluid and mobile, and weather is a narrative that we wish to tell and retell across time, over and over again. Weather mobilises us to feel a connection to places and times, to remember seasons past and to feel the delicious anticipation of future weather. The impending balmy summer nights, whilst trudging through cold winter rain, is both a remembrance of the past (summers) and anticipation of the future, about what will be when summer rolls around again. Or perhaps anticipation is more immediate, still sloshing through the rain, while thinking about being wrapped up in front of a cosy fire, dry from the fierce, lashing weather outside. Each of us is like a small world, infused with and by weather, as Mark Tredinnick explains, 'I harbour weather; I am made of it. And this helps me understand the world, and all the weather it suffers, how the world and all of us within it are weathered, without end' (2013: 14). Weather is always on the move and our feelings about interaction with the weather fluctuate and are deeply and viscerally felt (*and this helps me understand the world, and all the weather it suffers*). The weather, too, impacts on bodily forms such that the rain soaks the skin; goosebumps rise when a breeze flows on bare arms; we tan, freckle and wrinkle under the sun's rays, our skin becoming darker; and chills are able to seep into bones, cramping muscles.

As Tredinnick (2013: 12) eloquently writes, 'we live inside the weather. There's no escaping it: it is how the world speaks to us; it tempers and colours all our days and nights. It clothes us; it decorates and articulates the places where we live'. Thus, weather is sheer movement. That is, experiences of the weather cannot be understood outside of weather's constant movement (Vannini et al., 2012). For farmers, weather tempers and permeates lived experiences and livelihood, it also stimulates a cascade of mobilities across farms. Crops are sown or harvested in favourable conditions, and livestock are moved when pasture is too wet, too dry, or cows may be pinned back from too much lush grass by electric tape. Production practices, then, are formulated around weather.

In this chapter, I use the concept of dwelling to interrogate farming in Aotearoa and how it touchpoints with weather mobilities. Dwelling, as a place-making process is emergent, lived, experienced and in motion, as is the weather. As Vannini et al. (2012: 364) point out: 'weather movements over time shape, not only the material features of place but also provide affordances for the human engagement of place.' As such, the farmers in this research practically engage with weather to formulate dairy production in the Waikato region. The Waikato region, located in the North Island of Aotearoa New Zealand (hereafter, Aotearoa), is predominantly known for its dairy production with international milk-producing industry giant Fonterra's headquarters nestled into its rich floodplain soil. It is no accident that the dairy industry has flourished in the Waikato area, it has been afforded by the temperate weather and landscape conditions ideal for high-production grass growth. Similarly, milk production mobilises a huge legion of tankers, a common sight on rural and main roads. Milk transportation to hubs (dairy factories) has shaped the economies and settlement patterns of many Waikato townships. Therefore, dwelling is not a mere place of being, but a 'taskscape' (Ingold, 2000: 197), a mutually interlocking set of tasks that 'come into being through movement'.

Weather is highly important for livestock health and well-being, as dairy farming in Aotearoa is predominantly conducted outdoors. Cows in the Waikato region are seldom inside but kept on pasture grass all year round. The surplus of grass in spring and summer is cut for hay or made into silage or haylage and fed out in the months when grass growth is insufficient. The lactation of cows, too, is timed to make the best use of grass growth, 'with the spring flush of milk production coinciding with the highest rate of grass growth' (Blunden et al., 1997: 1765). Weather, then, is central to farming life, with droughts, steady weather or floods directly impacting farm dwelling. Farming in Aotearoa, put simply, is at the mercy of weather – good and bad production rates rely heavily on successful, rich grass growth which are directly correlated to climatic conditions.

Weathering and dwelling are related in the farming context. Weathering can be conceptualised as a wearing down, enduring ('weathering the storm', for example), eroding but also as an intimate knowing. A weather-beaten face is one that is enfolded and inscribed by life and the weather – it is weathered. In this context, weathering involves a transcorporeal stretching between past, present and future, which may be utilised in order to reimagine bodies (farmers, cows) as archives of climate and as making future climates possible (Neimanis and Walker, 2014: 558). Relatedly, post-humanist performativity is incorporated to denote the human and nonhuman entanglements of climate and weather. For farming, cows' bodily rhythms are orchestrated by weather; calving, milk production and breaks from milk production are seasonally linked with temperate climates favoured over harsh extremes. Farmers too are mobilised by the rhythms of livestock needs, such as the well-worn understanding among dairy farmers that they must wake in the dark cool hours of early morning to milk the cows. All bodies have their own rhythms of weathering; however, these rhythms can and are mobilised in different ways.

Defining a climate, including what is deemed to be a usual pattern or not, has been as much about observation and cultural interpretation as meteorology and climate processes (Hulme, 2008). Today's understandings of climate, and weather, are increasingly large scale and abstract, constructed through conceptualising climate change as a global phenomenon produced via oceans, human processes and weather systems. Weather and climate, however, are also shaped by demands and values that are resolutely local (Brace and Geoghegan, 2011; Hulme, 2008). The ways in which people experience and talk about the weather is rich and nuanced; the weather is sensed and comprehended and draws significance from meteorological processes that are transcorporeal and 'thick' (Irigaray, 1992; Neimanis and Walker, 2014). For example, my mother talked intimately about the limestone bluffs that filled her view from the farm cottage in which we grew up. The bluffs changed with each passing of a cloud, striations of light playing across an indifferent chalky face. Accordingly, people develop emotional attachments to the weather; they create an archive of weather experiences that binds them to places, landscape and identity. In this context, Tim Edensor's (2012, 2017a, 2017b) work on light and illumination explains the manifold effects of light and how these shape everyday experience. Light is refracted differently in changing weather patterns, and thus, the landscape is experienced in diverse ways when mediated by the artful staging of light across land and water (Edensor, 2017a). Those limestone cliffs, weathered through time, bind my mother to her 'home' and the farm in Aotearoa, in a similar way that dust and melted heatwaves permeate a sense of Australian-ness to Tredinnick (2013).

Experiences of weather, however, move beyond the individual and are particularly valuable as keys to deciphering large-scale social processes (Vannini et al., 2012: 363). Weather talk permeates the idea of dwelling, a place-making process, always in motion like the weather. The 'weather-world' (Ingold, 2010) is a concept that allows exploration of the ways in which people experience living the weather. For transcorporeality, as noted above, our minds and bodies are inextricably connected to and interwoven with the fabric of the world, so that mind/ body/weather-world are closely braided and mutually influencing. This movement of weather across bodies forms a part of our intimate subjectivity, a 'mingling' (Ingold, 2007) with the elements.

The difference between weather and climate is a difference of time and scale (Neimanis and Walker, 2014). I begin by discussing how mobilities, dwelling and weather come together in both time and scale in the farming context. Temporality is considered by overviewing the ways in which farming has mobilised landscapes over time. Farming as an economic practice has colonised parts of Aotearoa, profoundly changing landforms via intensification of practices. Scale, too, is important to the contemplation of weather and climate. Global climate change is abstracted in different ways but is intimately connected to the local scale. Farmers tread between the local and the global when mobilising their weather knowledge for milk production on the global market. In short, the weather mobilises landscapes, people and cows in varying ways. These transcorporeal connections are examined by utilising dwelling as a concept, paying particular attention to the patterns of weather that reframe the mobilities of cows, people and land.

Methodological note

The empirical data drawn on in this chapter was gathered from a small year-long pilot project with several Herd-Owning Sharemilkers (HOSM) who live in the Waikato region. Sharemilking is a largely unique agrarian practice in Aotearoa and is embedded in the notion of mobility. Sharemilkers typically own the livestock and machinery, but not the land, entering into a contractual relationship with landowners on a profit-sharing basis. The contract nature of sharemilking stimulates a multitude of mobilities as farmers move their stock and equipment to a new contract at a minimum period of three (milking) seasons (see Adams-Hutcheson, 2017a, 2017b). Interviews were conducted with HOSM in their homes and separately out around the farms, to follow Büscher and Urry's (2009: 103) contention that 'by immersing themselves in the fleeting, multi-sensory, distributed, mobile and multiple, yet local, practical and ordered making of social and material realities, researchers have gained an understanding of movement'– in this context, of dwelling in a weather-world. As such, interviews were conducted in all kinds of weather and had varying impacts on the researcher, participants and cows (for detail see Adams-Hutcheson, 2017a).

Farming and weather over time

Place, undoubtedly, plays a central role in influencing and shaping weather and people (and their memories of it). Aotearoa, like everywhere, has been shaped, moved and moulded by the weather, both ecologically and socially. Weather and climate have mobilised people, industry, economies and livelihoods. It is important to impart that what appears as fixed forms in the landscape, unmoving and inert, unless acted upon from outside, is itself in motion (Ingold, 1993). This sense of motion is dependent on a timescale that is both immediate and 'immeasurably slower and more majestic' than that in which human activities are normally conducted. For instance, Ingold (1993) notes that when geological time is speeded up, we see glaciers flow like rivers; solid rocks bend, buckle and flow; as well as trees flex. Similarly, weather is often depicted in time-lapse film to capture the dramatic flow of elements such as storm clouds rolling across the landscape. A consideration of dwelling as movement (e.g. Mason, 2016; Vannini et al., 2012) and weather-worlds as always in motion (Ingold, 2010) coheres with Ingold's (2010: S122) insight that 'the experience of weather lies at the root of our moods and motivations', our movements and living spaces, as well as shaping landscapes and identity. Weather movements are able to shape the material features of place.

Dwelling is about the rich intimate ongoing togetherness of being and things which make up nature and places (Cloke and Jones, 2001: 651), and in the Waikato region, nature and culture are drawn closely together. The Waikato region, cut through by a large river, which carries the same name and runs for 425 kilometres, has ideal conditions for dairy farming. The volcanic soils are relatively free-draining and rich, essential for verdant grass growth; the rainfall is ample; winters are mild; and the softly undulating green hills are deemed to be particularly appropriate for dairy cows. The colonial utilitarian attitude towards land saw the rise of the dairy industry, intensification of farming and global milk production. The transformation of land over time is considered here; the Waikato region moved from being a wetland and densely forested basin, important and rich in food sources for Māori, to being cleared, burnt off and drained as many trees fell to the axe (Dench, 2011). Thus, the land has been mobilised over time as colonists turned to different models of climate processes to expedite land settlement and gain economic benefits.

If colonisation meant taking land for the settler project, it also involved promoting scientific weather knowledge and often erased alternatives. Māori expressed a deeply embedded understanding of weather and climate in their oral traditions. Local weather and extreme events were passed down through generations that helped Māori adapt to climate variability and make decisions about gaining a livelihood from both ocean- and land-based resources (King, Skipper and Tawhai, 2008). Not only did settlers often overwrite Indigenous conceptualisations of weather and climate with their own but they also used climate theories to limit or justify industry in particular places. Peter Holland and Jim Williams have noted that some Māori did exchange extensive weather lore with Europeans. European conceptions of science, however, still came to dominate local understandings of climate and weather (cf. O'Gorman et al., 2016: 897), as well as dominate the landscape with agricultural practices.

Colonial weather and climate knowledge have had an essential and fundamentally practical purpose – to ensure settlement's success. Settler economies, such as those in Aotearoa, depend on the weather in regular patterns and rhythms to sustain primary productive economies including farming (O'Gorman et al., 2016: 898), mobilising the dairy industry in favourable areas such as the Waikato region. Mobility, then, is always in relation to someone or something else, and in doing so, helps to connect or perhaps differentiate people and places. This can also foreground an appreciation of the power and politics bound up in mobility and immobility (Adey, 2006; Cresswell, 2006). The weather has found its place in mobilities; accordingly, I seek to explore the mobilities of weather as political, personal, relational and contentious.

Mobilising local and global weather scales

Climate change is undoubtedly political and contentious. Bodily, socioeconomic, historical and geopolitical differences complicate how we weather the world. Indeed, the weather is understood to be shaped by demands that are at once local, physical, political, historical, social and economic. As Georgina Endfield and Sam Randalls have argued:

> climate is a philosophical and political category as much as it is a material category, one that was deployed by a diversity of actors in changing and sometimes conflicting ways throughout the British Empire and beyond.
>
> (2015: 21)

Climate (and climate change) happens on a global scale over long time periods, while weather is conceptualised as more immediate, such as, what is happening outside the window right now. The distinction between climate and weather, it may be argued, means that climate and its changing patterns are deemed to be abstract – an abstraction that is happening elsewhere to other people (Brace and Geoghegan, 2011; Duxbury, 2010; Hulme, 2007, 2008; Neimanis and Walker, 2014; Slocum, 2004). Equally climate science has produced vast quantities of knowledge that is also abstracted through statistics and models. The complexity of this data has effectively alienated local response, cultural contexts, Indigenous forms of knowledge, lived experiences and participative practices. Climate change then travels across scales unfettered from cultural anchors with little local resonance. The construction of universalised indicators of change (the climate is warming; the seas are rising) strips them of their human touchpoints, affects and cultural meanings (Hulme, 2008). For example, there has been a warming climate trend across Aotearoa of approximately 1.1 degrees Celsius between 1900 and 2009 (Kalaugher et al., 2017: 53). This rather abstract set of numbers is devoid of the pressures and heartache of getting dairy cows through a drought. For example, a sharemilker, Alex stated that,

> You can't make light of the drought when the cows are mournfully bellowing and lowing over the fence [because they are hungry], townies [urban people] can't understand how dreadful it is, there's no escape.
>
> (Interview 16 November 2016)

Abstracted weather statistics render the cow and farmer bodies and relational mobilities as being absent in the discussion of droughts.

Climatic fluctuation or discordant rhythms of weather 'patterns' demand innovative responses from farmers to environmental risks that emerge in relation to (or in anticipation of) modifications arising from changing weather 'within a relational context that may include the places that people live, their histories, daily lives, cultures or values' (Slocum, 2004: 416).

Another sharemilker, Greg, mentioned that, 'We've had so many droughts in the recent past, I've had to modify everything I do, including not having as many cows on the place' (Interview 28 September 2016). Recent scholarship on the significance of locality in environmental experiences demonstrates that human and animal resilience to long-term climate shifts, and short and severe weather events such as drought or flooding, depends on many variables, from an individual capital to water access to prevailing government policy.

Climate change is fundamental to mobilities scholarship, where investigating the changing climate draws attention to climate migration and human mobility as a form of adaptation. Climate adaptation to local weather events weaves studies that bring climate and the weather together (Brace and Geoghegan, 2011; Endfield, 2011; Hulme, 2008). The structural impact of a climate event on mobility is therefore a question requiring preliminary subjective analysis. As Parsons submits, only by 'understanding precisely what a flood, drought, or storm means to those who experience it is it possible to meaningfully interpret its impact on mobility' (2019: 682). For farmers, adaptation to climate change is part of a continuous iterative process of adapting to changes which stimulate and mobilise localised farm movements in different ways that affect farm management (Kalaugher et al., 2017). Individual farm-stocking rates ebb and flow, irrigation is mobilised, supplementary feed is likewise mobilised to keep production rates steady and aid animal health and welfare. These changes in weather are felt and responded to at the farm level. Therefore, weather stimulates and exacerbates different scales of mobility, and ultimately ties together the weather and patterns of climate.

Farming as dwelling in weather

In dwelling in a weather-world, farmers adapt to unique weather-places in localised ways. Farmers are intimately in tune with the circadian rhythms of livestock and crops and possess extensive knowledge of weather patterns that are enmeshed with their everyday lives. Farmers cultivate a relationship to place that is attuned to the rhythm of seasons and acknowledges the temporality of landscape (Ingold, 1993). Indeed, farmers have organised their spaces of agriculture through historical understandings of the weather, the rhythms of nature to which they have become accustomed over time. In recent years, meteorological science has made enormous progress in predicting climate but there has been disagreement on weather data ownership and lack of agreement for countries to work together and set standards. In the past, weather data was freely shared, but increasing costs and declining budgets have threatened the freedom of availability of weather information.

Underpinning the freedom or not of availability of weather information, for Aotearoa in particular, were different philosophies about the role of a governmentally funded meteorological services, which has seen the development of a partially privatised weather service. The New Zealand Met-Service, a government owned corporation is seen as a successful business delivering widely available weather services (National Research Council, 2003), and has thereby diminished the possibility of a deregulated weather market. Henry (2015) has highlighted how gaps in local weather prediction has prompted the rise of private forecasters, who offer farmers a subscription service in competition with public meteorological services. This has provoked a series of often-acrimonious debates between public meteorologists and private forecasters about the reliability of the former and the credibility of the latter (O'Gorman et al., 2016). Most of the farmers I spoke with reinforced the importance of the larger weather data providers such as Met-Service, as they felt that too much is at stake for their livelihood to risk misinformation. Farms are quite simply mobilised by the weather. The importance of weather data was obvious within interviews and its impact can be huge, as farmer Sven outlined:

> We knew that rain was coming. I subscribe to the monthly outlook from Met-Service. We designed our whole spring rotation around a big weather system relayed by Met-Service, we had such a good winter. I look at the weather three times a day, it changes constantly and farming is like a symphony that works in harmony around the weather, literally everything is orchestrated by it.
>
> (Interview 9 November 2016)

Farming decisions about how many stocks are appropriate for the amount of grass produced (grass growth to balance production), soil health and animal health rely on accurate weather data. Decisions about balancing soil nutrients to enhance production are based on technical (weather apps; websites; national/local weather data platforms) and non-technical information, often in the form of farm-specific weather records (weather diary entries and field-notes). For example, Jeff related that,

> Last year there was a drought event forecast under the El Niño Southern Oscillation (ENSO) swing and we were supposed to get a real dry patch. We changed our farming right away. Right from August [winter], we pushed grass forward and kept the round [of cow rotation] really slow. We just kept building up the feed wedge so we would have plenty of feed available come February/March [summer].
>
> (Interview 17 November 2016)

Farmers mobilise the farm and animals to adjust to a possible string of drought events because such a trend would matter in practical ways. For instance, a farmer may buy extra feed, or as in the above quote, build up resources of feed on the farm. Other steps may be taken, such as to perhaps cull weaker cows in the herd earlier during the onset period of a drought or cut personal spending to finance new water storage facilities. Dwelling offers insights into how (non)human actants are embedded in landscapes and places, how nature and culture are bound together in place (Cloke and Jones, 2001: 650). These formations invariably have a link to temporality but also scale where the local farm response and bigger climatic patterns converge. Dwelling is an emergent and fluid process, which constantly shapes the ongoing togetherness of beings (farmers, cows) and things (crops, grass, feed-pads) that make up landscapes and places. Ambiguity in the weather is perceived when events do not occur as expected. The string of droughts mentioned earlier by Greg (above) may be interpreted as an increase in the normal incidence of drought over the previous records kept for a particular farm. Farmers may choose to decide that these are isolated events or perhaps signal a change in climatic patterns, drawing the scale of weather and climate close (Hulme, 2008). Attuned weather data at the local level, however, helps farmers to make their decisions as early as possible to mitigate negative impact on cow health, well-being and production rates.

Mobilising weather talk

The importance of overall weather patterns and seasonal fluctuations of wet and dry periods in farming cannot be understated. Daily and seasonal farm management evaluations are largely directed by weather conditions. Weather forecast information is constantly disseminated with farmers often interpreting weather at highly specific levels such as a consideration of the millimetres of rain that are expected to fall (see NIWA FarmMet, 2016). The interpretation of weather is also culturally engrained, where 'weather talk' is expected, particularly in the farming community. Weather came up in all my interviews. For example,

> I ask my friends over the hill, how much rain did you get? We all talk in mils [millimetres] of rain constantly. I don't know anyone who doesn't have a rain gauge, we check it every day.
>
> (Brent, Interview 20 November 2016)

> We had some overseas visitors and asked them how they found New Zealanders? They said, 'They're nice but they talk SO much about the weather!'
>
> (Nicola, Interview 20 November 2016)

> I take it for granted, my whole life I've talked about x amount of mils [of rain] this day and that day. It's such a necessary part of our farming life.
>
> (Scott, Interview 17 December 2016)

Talking about the weather is at first sharing information, and then responding to its minutiae details. In short, what is divulged about weather on the farm carries with it a corresponding action. Farmers notice that when the overall weather patterns on their farm change, they need to plan ahead for production levels, soil health and animal health. Thus, by integrating historical records and current day-to-day data to predict the next season's weather, they draw on past, present and future simultaneously in their analysis of climatic conditions. Farmers build up and create their own intimate archives (Neimanis and Walker, 2014) of knowledge about temperature, precipitation and the onset of different seasons. In this way, the HOSM accounts of the climate are a mingling of place, personal history, daily life and values. Farming binds life to the weather, where one is immersed in the incessant movements of wind and rain – indeed in all elements. Farmer Brent discusses his view, stating that:

> When you're farming, you've always got to look down the track [thinking ahead]. You have to see past the drought you're in or see past the wet, muddy spring weather and know that it will come right. You've just got to get into your wet weather gear, when it's pouring down, and get out there in it and farm the place, look after the cows.
>
> (Interview 20 November 2016)

It is through the weather that the farming community experiences daily, personal visceral encounters (Geoghegan and Leyson, 2012). The seasonal rituals of planting, the moving of stock from place to place, building-up resources for wet or dry periods, and the immobilities of stock during winter produce an intimate, mobile, rhythmic imaginary. The interviews with HOSM were peppered with dialogue on how cows are moved to drier parts of the farms for calving during winter and may stand on feed-pads so as not to damage pasture or create excess mud. In the spring flush of grass, mobility is restricted to utilise pasture, and during dry summer months, the cows range further in the fields and have access to shade trees. These cow rhythms and mobilities are relational to the weather, as cows, too, dwell in a weather-world. One of the conceptual advantages of thinking of animals in terms of movement and dwelling in the weather-world is that it provides a concrete way of grappling with key ontological questions concerning animal agency, as Brent states:

> Well, it's been so wet, obviously. You know? It's the wettest season in 20 years that we have had. Too much rain does cause you a lot of, constant, change of plans. Normally I farm a day or two out, maybe even a week ahead, but those plans change the wetter it gets. Sometimes you're shifting cows right at the last minute, that's how crucial the weather can be. The cows are restless too, they don't like the wind and driving rain and often break [through the fence]. They also don't like being messed with, you know, changed suddenly – it disrupts their routines and they sulk.
>
> (Interview 20 November 2016)

By thinking of animals as modalities of movement, or thinking about their mobilities, there is the potential to move away from the 'mere extension to animals of the humanist and anthropocentric conceptions of agency still typical of much of social science', and to move, instead, towards a more collective, distributive and relational conception of animal dwelling (Nimmo, 2011: 71). Animal agency and dwelling may allow an extension of learning about a relational weathered existence together.

Conclusion

In this chapter, I have approached my inquiry of HOSM and their intimate relationship to weather and climate through integrating the concept of dwelling. Dwelling is also extended by mobilities, paying close attention to how the weather influences movement of human and nonhuman entities on farms. Research that focuses on movement, connectivity and experience means that HOSM lives were examined *in situ* to facilitate a deeper understanding of the multisensory daily tasks of farming, which includes cows, farms, farmers and different climatic conditions.

Weather is mobile and inescapable, we are always *in* it (Ingold, 1993); in the farming context, it is also highly political. This political aspect comes in not only because of the climate change context in farming but also because of uneven individual and collective farming experiences of weather. Weather, as this chapter has explored, is also historical and bound to place-making and identity. For Aotearoa, the weather is colonised, where settler knowledge was foregrounded over Indigenous ways of understanding the weather-world. Māori demonstrated an intimate understanding of weather and climate across their tribal lands. Oral recordings offer insights into how Māori adjusted to past events with acute awareness of local weather and climate phenomena (King, Skipper and Tawhai, 2008), but these were often sidelined in the colonial project of settlement. Weather was of supreme importance in moulding and mobilising the landscape of Aotearoa towards agricultural economies. Over time, settlers sought ways to understand, predict and manipulate climates to their advantage instating agricultural practices (such as farming) into local landscapes. Thus, in legitimating Western agricultural practices, meteorology was important as the 'science of the Empire'.

Literature on farming and the climate has investigated local responses, mostly to climate change and farmers' intimate knowledge of their localities (Geoghegan and Leyson, 2012; Holloway, 1999; Riley, 2011). I considered the usefulness of dwelling for weaving together climate and weather by focusing on experiences of weather at the farming level. The farmers' experiences in this chapter are crucial in drawing together weather and climate, that is, the local experiences of weather patterns and their corresponding stimulation of farm mobilities, farmer mobilities and cow mobilities. In discussing dwelling, it is important to acknowledge the ability to pay attention to the weather, cows, farmers, crops and the landscape which are relational and in-fused with movement. In these relationships, the weather is transcorporeal. Excerpts from interviews were used to explain how weather events stimulate multiple movements on the farms, infuse day-to-day and future planning and how cows are impacted by the weather, for example, being unsettled in wet and stormy conditions.

HOSM mobilities offer a useful platform to think about dwelling in a weather-world, but also how the weather precipitates movement of humans, animals and nonhuman entities such as crops and machinery on farms. Indeed, farming in the Waikato is orchestrated by the weather and movement. By focusing on the experiences of HOSM, this study extended this weather/mobilities intersection to include the intimate politics of time, landscape (scale) and identity.

References

Adams-Hutcheson, G (2017a). Farming in the troposphere: drawing together affective atmospheres and elemental geographies. *Social & Cultural Geography,* 20(7): 1004–1023.

Adams-Hutcheson, G (2017b). Mobilising research ethics: two examples from Aotearoa New Zealand. *New Zealand Geographer,* 73(2): 87–96.

Adey, P (2006). If mobility is everything then it is nothing: towards a relational politics of (im)mobilities. *Mobilities,* 1(1): 75–94.

Blunden G, Moran W and Bradly, A (1997). 'Archaic' relations of production in modern agricultural systems: the example of sharemilking in New Zealand. *Environment and Planning A,* 29(10): 1759–76.

Brace, C and Geoghegan, H (2011). Human geographies of climate change: landscape, temporality, and lay knowledges. *Progress in Human Geography,* 35(3): 284–302.

Büscher, M and Urry, J (2009). Mobile methods and the empirical. *European Journal of Social Theory,* 12(1): 99–116.

Cloke, P and Jones, O (2001). Dwelling, place and landscape: an orchard in Somerset. *Environment and Planning A,* 33(4): 649–666.

Cresswell, T (2006). *On the move: mobility in the modern western world.* London: Routledge.

Dench, S (2011). Invading the Waikato: a post-colonial review. *New Zealand Journal of History,* 45(1): 33–49.

Duxbury, L (2010). A change in the climate: new interpretations and perceptions of climate change through artistic interventions and representations. *Weather, Climate and Society,* 2: 294–299.

Edensor, T (2012). Illuminated atmospheres: anticipating and reproducing the flow of affective experience in Blackpool. *Environment and Planning D: Society and Space,* 30(6): 1103–1122.

Edensor, T (2017a). Seeing with light and landscape: a walk around Stanton Moor. *Landscape Research,* 42(6): 616–633.

Edensor, T (2017b). *From light to dark: daylight, illumination, and gloom.* Minneapolis, MA: University of Minnesota Press.

Endfield, G (2011). Reculturing and particularizing climate discourses: weather, identity, and the work of Gordon Manley. *Osiris*, 26(1): 142–162.

Endfield, G and Randalls, S (2015). Climate and empire. In Beattie, J, Melillo, E, O'Gorman, E (eds.), *Eco-cultural networks and the British Empire: new views on environmental history*. London: Bloomsbury, pp. 21– 43.

Geoghegan, H and Leyson, C (2012). On climate change and cultural geography: farming on the Lizard Peninsula, Cornwall, UK. *Climatic Change,* 113(1): 55–66.

Henry, M (2015). 'Inspired divination': mapping the boundaries of meteorological credibility in New Zealand, 1920–1939. *Journal of Historical Geography*, 50: 66–75.

Holloway, L (1999). Understanding climate change and farming: scientific and farmers' constructions of 'global warming' in relation to agriculture. *Environment and Planning A*, 31(11): 2017–2032.

Hulme, M (2007). Viewpoint: understanding climate change – the power and the limit of science. *Weather*, 62(9): 243–244.

Hulme, M (2008). Geographical work at the boundaries of climate change. *Transactions of the Institute of British Geographers,* 33(1): 5–11.

Ingold, T (1993). The temporality of the landscape. *World Archaeology,* 25(2): 152–174.

Ingold, T (2000). *The perception of the environment: essays in livelihood, dwelling and skill.* London: Routledge.

Ingold, T (2007). Earth, sky, wind, and weather. *Journal of the Royal Anthropological Institute*, 13(S1): S19–S38.

Ingold, T (2010). Footprints through the weather-world: walking, breathing, knowing. *Journal of the Royal Anthropological Institute*, 16(S1): S21–S39.

Irigaray, L (1992). *Elemental passions* [Trans. Collie, J and Still, J] New York: Routledge.

Kalaugher, E, Beukes, P, Bornman, JF, Clark, A and Campbell, D (2017). Modelling farm-level adaptation of temperate, pasture-based dairy farms to climate change. *Agricultural Systems,* 153: 53–68.

King, DNT, Skipper, A and Tawhai, WB (2008). Māori environmental knowledge of local weather and climate change in Aotearoa – New Zealand. *Climatic Change,* 90(385): 385–409.

Mason, J (ed.) (2016). *Living the weather: voices from the Calder Valley.* Manchester, NH: Morgan Centre for Everyday Lives.

National Research Council (2003). *Fair weather: effective partnership in weather and climate services.* Washington, DC: National Academies Press.

Neimanis, A and Walker, RL (2014). Weathering: climate change and the 'thick time' of transcorporeality. *Hypatia*, 29(3): 558–575.

Nimmo, R (2011). Bovine mobilities and vital movements: flows of milk, mediation and animal agency. In Bull, J (ed.), *Animal movements, moving animals: essays on direction, velocity and agency in humanimal encounters.* Uppsala: Centre for Gender Research, Uppsala University, pp. 57–74.

NIWA FarmMet (2016). Farm operational decisions based on powerful web-based subscription weather forecasting. Retrieved November 9, 2016, from https://farmmet.niwa.co.nz/#/about/package/FarmMet.

O'Gorman, E, Beattie, J and Henry, M (2016). Histories of climate, science, and colonization in Australia and New Zealand, 1800–1945. *Wiley Interdisciplinary Reviews: Climate Change*, 7(6): 893–909.

Parsons, L (2019). Structuring the emotional landscape of climate change migration: towards climate mobilities in geography. *Progress in Human Geography*, 43(4): 670–690.

Riley, M (2011). 'Letting them go' – agricultural retirement and human-livestock relations. *Geoforum*, 42(1): 16–27.

Slocum, R (2004). Polar bears and energy-efficient lightbulbs: strategies to bring climate change home. *Environment and Planning D: Society and Space*, 22(3): 413–438.

Tredinnick, M (2013). The weather of who we are: an intimate essay on the weather, the self and Australianness. *World Literature Today*, 87(1): 12–15.

Vannini, P, Waskul, D, Gottschalk, S and Ellis-Newstead, T (2012). Making sense of the weather: dwelling and weathering on Canada's rain coast. *Space and Culture*, 15(4): 361–380.

16 Nuclear warfare and weather (im)mobilities

From mushroom clouds to fallout

Becky Alexis-Martin

At 5:29 am on 16th July 1945, humans created the first-ever atomic mushroom cloud in history when they detonated the Trinity atomic bomb test in Alamogordo, USA. The atomic mushroom cloud is a phenomenon of humanity's techno-military obsessions, and one that retains its aesthetic entanglements with Cold War power. It is superficially manifest as an ephemeral event but, paradoxically, is a tactile and lingering marker of scientific progress and unprecedented power.

Under scientific scrutiny at the Trinity test site, this original monstrous pyrocumulus cloud ballooned and billowed as it rose, forming a toroidal vortex as the surrounding air cooled. Its core was incandescent as it rapidly morphed into its unnatural form. Milliseconds passed as a battalion of specially invented Rapatronic high-speed cameras documented its every expansion – in the name of science, warfare and progress (Rosenthal, 1991). Airburst nuclear weapons are atmospheric explosions and, as such, they judder local weather conditions into something phenomenal and strange. The Trinity Test was expansively sensational, operating beyond the aesthetic of the dynamic sublime as weapon scientists' bodies 'registered the power of the bomb' (Kant, 1987; Nye, 1994; Masco, 2008: 352). Thus, weapon scientists and spectators alike were caught up within the transformations of the first *nuclear weather-world*, one of localised immersion in the generative fluxes of the nuclear medium – in the wind, heat, sound and sights of the atomic blast (Merleau-Ponty, 1963; Ingold, 2007). Brigadier General Thomas F. Farrell was one of ten men selected to observe the first nuclear weapon test. His description paints a vivid experiential portrait of the early mushroom cloud. He was awestruck by its exquisite beauty and fierce light-play: '(I)t was golden, purple, violet, grey and blue. It lighted every peak, crevasse and ridge of the nearby mountain range with a beauty that cannot be described but must be seen to be imagined' (*Washington Post*, 1945). Lightning flashed. The smoke trails of scientist's sounding rockets transcribed the pathways of invisible yet tangible shock waves across the sky.

This first nuclear weapon detonation marked the dawn of a freshly *atomic Anthropocene*, creating localised air temperatures so hot that silica sand seared into glassy trinitite, fossilising the heat of the blast into puzzling traces for future archaeologists. The event was an initiation ceremony to a new lethal and otherworldly set of weather conditions, as fallout was swept into the atmosphere by convection and prevailing winds only to pepper down across the New Mexico desert. Thus, the Trinity Test was the origin of *nuclear weather mobilities*, changing our human relationship with atmospheric radiation. It presented a new set of intimate and immediate relationships and conditions with *unnatural weather* for those who pass through, migrate from or otherwise experience weather induced by nuclear warfare. Beyond the Trinity Test, a diverse array of mobilities has since emerged that pertains to the nuclear, from the quotidian to the fantastic, from accident to warfare, from the temporary to the permanent, and from communities to the material pathways of atmospheric nuclear detritus itself (Cresswell, 2011; Rush-Cooper, 2013; Alexis-Martin, 2015; Davies, 2015; Alexis-Martin and Davies, 2017, Alexis-Martin, 2019a).

A commonality that runs through the geography of nuclear weather is the dichotomy of nuclear (in)visibility and the human body's limitations in perceiving radioactivity itself. Beyond the intensely experiential blast of a nuclear accident, weapons test or act of war, human perception of the nuclear is reliant upon second-order observation of 'nuclear weather' to determine the presence of the nuclear sublime, whether this is the murky precipitate of radioactive fallout or the metallic tang of ash. This (in)visibility is a definitive feature of nuclear weather mobilities, as individuals and communities move across space and time to evacuate – or perhaps unknowingly undertake daily routines – within the diffuse and fuzzy extent of any radioactive plume. The (in)visibility of nuclear weather produces an unknown element of how and where people move and react, revealing disparities of mobility across horizons and verticalities of space-time. Those who are privileged to possess political and spatiotemporal knowledge of the bomb have the ability to select alternative and potentially safer mobilities during nuclear weather events, while those without this knowledge or ability to grasp the extent of this new form of nuclear weather are unable to make these choices. Thus, the human outcomes of weather produced by nuclear warfare are possessed of a 'power geometry', whereby the mobilities of some are very much dependent upon the (im)mobilities of others for nuclear warfare to be effective (Massey, 1993). In fact, the purpose of nuclear warfare could be described in these terms – to blast a freshly uneven power geometry into existence, enhancing the sociopolitical mobilities of nuclear warmongers while destroying those on the ground.

This chapter interrogates the processes, people and phenomena that surround *nuclear warfare weather mobilities* through select examples during the Hiroshima and Nagasaki bombings, and the Cold War nuclear weapon tests, from the imprecise choreography of mushroom clouds to the forced and inherently necropolitical mobilities of Indigenous people (Mbembe, 2019). When we consider the overarching mobilities pertaining to nuclear weather, for the general population they can be articulated in relation to the nuclear event as 'responses to unexpected, unexplained...events or occurrences' (Adey et al., 2014:14). As a human-made irregular warfare event, nuclear warfare weather can be anticipated and explained as an intentional and knowing mobility-inducing or inhibiting scenario by the nuclear warfare actor that produces and exposes uneven, multi-scalar and sometimes lethal power relationships from the individual to the international, through these (im)mobilities (Jones and Smith, 2015).

Under a mushroom cloud

On the 6th and the 9th of August 1945, respectively, the Japanese cities of Hiroshima and Nagasaki were destroyed in nuclear attacks by the USA. This was the first time that nuclear weapons were used in warfare – and this action was facilitated not just by bombastic displays of new technology in the New Mexico desert but by a culture of American anti-Japanese racism. American propaganda had dehumanised the Japanese people by presenting them as expendable vermin, thus enabling an 'ethical' atomic bombing of Hiroshima and Nagasaki (Aoki, 1996). For the residents of Hiroshima and Nagasaki, inescapable immobilities manifested at that split-second of detonation – as each city was consumed by fireball, blast and irradiation. Among those who survived, many suffered debilitating injuries and long-term life-limiting health issues (Alexis-Martin, 2019a). It was a triple-exposure of immobility, as the transport networks of both Hiroshima and Nagasaki were also destroyed (Matsunari and Yoshimoto, 2013).

The testimonies of the Hibakusha (the survivors of the atomic explosions, a term meaning 'the exposed' and initially loaded with stigma, but which has become an honorific over time) of Hiroshima and Nagasaki give insights into the lived experiences of such rapid destruction while under a mushroom cloud. Both mobilities and weather conditions feature prominently within their descriptions of living through nuclear warfare. For example, Mr Takato Michishita (2017) described his personal experience of surviving Nagasaki:

> Everything turned white. We were too stunned to move, for about 10 minutes. When we finally crawled out from under the tatami mat, there was glass everywhere, and tiny bits of dust and debris floating in the air. The once clear blue sky had turned into an inky shade of purple and grey. We rushed home and found my sister – she was shell-shocked, but fine.

Here, we learn of the instantly shattered and diffracted mobilities that arise within the mushroom cloud. The stop–start reactivity of the motionless, and then the frantic survivor, fearfully seeking out their kin under ominous skies.

However, this is not the end of the story or perhaps even the beginning. To understand Hiroshima and Nagasaki through the lens of the mushroom cloud, we must also consider natural weather mobilities, specifically those that facilitate or prevent nuclear warfare. In the case of Nagasaki, there is a counter-story of another Japanese city that should have been the target had it not been for local atmospheric conditions. This city remains undisturbed by any atomic bomb and is known as 'Lucky Kokura'. When the Bockscar bomber arrived at Kokura on the morning of 9th August, the view of the city was obscured by a combination of natural haze – smoke from the recently bombed adjacent city of Yawata and also possibly from an intentional release of steam by the local electric power station (Bernstein, 1998; Wellerstein, 2015). After 45 minutes of gazing into the fug, the crew of Bockscar had to re-mobilise to their new destination, and Nagasaki's fate was sealed.

For Hiroshima, further weather-related consequences were yet to arise. On 17 September 1945, the city was deluged by the Makurazaki Typhoon, which caused an estimated 3,000 fatalities and severely damaged transport infrastructure (Hoshi et al., 1992). This caused a major setback to the redevelopment of the city. Dr Liebow's (1983) retrospective diary of his work in Hiroshima describes the event: 'During the great typhoon of September 17, a landslide roared down from the steep hills behind the hospital to the sea, crushing several buildings in its path'. The damage produced by the typhoon resulted in the destruction of the Ono Hospital and made both air and rail travel to Hiroshima impossible until 12 October 1945. These weather events compounded the immobilities caused by the destructive power of nuclear warfare.

Both local communities and international aid supported the gradual reconstruction of Hiroshima and Nagasaki. For Hiroshima, the city created a new five-year plan for land use that 'demarcated and re-organised' ruins (Nishii, 2019: 19). Land boundaries became nebulous and spaces were redrawn along adjusted lines, becoming new boundaries that changed people's daily lives (Nishii, 2019). The Peace City Reconstruction project added further complexity to local people's banal mobilities, as residents were relocated from damaged central districts to Yoshijima, a residential area outside of the city that offered few economic opportunities (Yui, 2003). Mobilities of everyday life were changed in many ways – from the daily navigation to and from work or school, to the relocation of entire localities.

The bomb forever morphed the mobilities of Hiroshima. International tourism emerged soon after the atomic bombing, with the blast zones offering an unparalleled spectacle for dark tourism. As early as March 1946, author Kotani Haruo called upon city officials to recognize 'the connection between peace and tourism, which will hasten the recovery of our city, the capital of the inland sea.' (Zwigenberg, 2016). This legacy means that the Hiroshima Atomic Bomb Museum now receives more than 1 million visitors as a site of international cultural importance and one of the earliest sites of 'dark tourism' (Schäfer, 2016). Thus, memorialisation and tourist mobilities peacefully co-exist.

It was a blast!

Trinity, Hiroshima and Nagasaki were the first examples of an expansive movement by nation states towards nuclear warfare during the early years of the Cold War. Aspirant global superpowers included the Soviet Union, the UK, France and China. These countries saw nuclear weapons as a way to regain geopolitical control, despite bankrupted national economies and diminishing empires (Alexis-Martin, 2019a). An arms race had quickly emerged between the USA and the Soviet Union by 29 August 1949, after the Soviet Union had tested its first atomic bomb, 'RDS-1',[1] at Semipalatinsk, Kazakhstan (Hawkins, 2013; Alexis-Martin and Davies, 2017). When nuclear weapon testing was undertaken during the Cold War, a paradox of mobilities arose. The processes required to produce and detonate a nuclear weapon created new spatialities and exclusion zones. Universally unequal mobilities arose for Indigenous communities, who found their lands suddenly co-opted, and the lives of many of the service people who were reluctantly conscripted before undertaking nuclear work were restricted and controlled.

Thus, nuclear weapon tests provoked unexpected forced mobilities and created pockets of oppression internationally. Nuclear weapons are 'the ultimate coloniser. Whiter than any white man that ever lived. The very heart of whiteness' (Roy, 1998: 205). Nuclear colonialism is the taking and destruction of other people's lands, natural resources and well-being for one's own benefit in the furtherance of nuclear development (Danielssons, 1986; Kahn, 2000). A nuclear imperialism–necropolitics nexus also emerged, whereby the geopolitical advantages of being a nuclear weapon possessor creates biopolitical disregard for the lives of people affected by nuclear weapon testing (Alexis-Martin, 2019b). This has been a universal feature of planning for nuclear weapons testing internationally, as 'empty' spaces have been sought out and dominated. For example, the UK, USA and France have created a combination of desert and island sites across colonial territories that collectively span the Indigenous populations of French Polynesia, the Marshall Islands and Kiribati, Algeria and Australia (Jacobs, 2013). Military and construction materials were shipped to these sites, tents were erected and barracks were constructed in pursuit of ultimate destruction.

For young British servicemen working on the Grapple series of tests on Kiritimati Atoll, also known as 'Christmas Island', their tour to test Britain's first hydrogen bomb (H-bomb) was described to them in fair-weather terms of 'paradise' and 'island life'[2] (Keown et al., 2018). However, their understanding of the nuclear weather sensorium that accompanied detonations was abject and inexplicable. A nuclear test veteran described his experiences of the Grapple Y test:

> We were told to stand with our backs to detonation and put our fists into our eye sockets, to cover our eyes. The flash was so bright that I could

see the bones in my hands. Like an X-ray, a million times brighter than the sun...My neck burned from the heat, and the wind! You couldn't imagine it.[3]

While the tests were meticulously scheduled, and anticipated with military precision and noisy countdowns, the phenomenal sights and sounds of the H-bomb were seared into their memories.

There was little warning before deployment to Kiritimati, due to the political sensitivity of the work. One nuclear veteran said, 'I was notified a week in advance. I was told to go to London airport with my entire kit bag ... we were men with overcoats and gloves, travelling to a tropical island' (Alexis-Martin, 2016). Some troops were deployed in other ways, such as the Royal Engineers who travelled en-masse to the island by boat with soldiers sharing facilities in the hull of ship, and sleeping in bunks that were three to four beds high.

> The whole unit went on a specially chartered train to Southampton, all thousand of us, then onto the boat and off we went...The ship got out of Southampton and then trundled off into Atlantic. Most people were seasick for the first week, until reaching the Bay of Biscay. I was lucky though, as I was on the top bunk and wasn't seasick. The first week there was no queue for breakfast because of the seasickness. It was great until people started to recover, then you can imagine the queues.

These conditions did not improve when the servicemen arrived at Kiritimati:

> The tents were very primitive. We slept on camp beds, with metal legs to clip in, about 6 to 8 inches off the ground. The tents were big, with at least ten people in each tent. We were provided better beds later on... It was the way things worked, occasionally the supplies arrived before the soldiers.

Weather was a constant hazard during the nuclear weapon tests, as servicemen spent most days travelling and working in tropical heat without sun protection. Their daily mobilities were determined by the coastline of Kiritimati and many servicemen experienced the tedium and loneliness of repetitive and banal work (Alexis-Martin, 2019a). While the conditions of being a nuclear test serviceman were challenging, many men were eventually able to travel home to their families and continue with their lives. For Indigenous communities that lived nearby and within nuclear weapon test sites, their sense of home and community was forever changed by nuclear weather.

Black mist

Nuclear weapon testing enforced unanticipated mobilities upon Indigenous communities. Spaces were territorialised, local civilians were often forced to leave their homes, and many eventually became nuclear refugees (Alexis-Martin, 2019a). The repercussions of this nuclear imperialism included trauma, health challenges and the loss of cultural ties, as communities tried to adapt to their unsuitable new conditions. For example, the Australian Aboriginal and Torres Strait Islander peoples of Maralinga tried to return home, only to discover barbed-wire fences surrounding their ancestral homelands (Reed and Stillman, 2010). Other communities, including the Kazakh nomads of Semipalatinsk, were left in place during Soviet Union nuclear weapons testing and now live with intergenerational genetic health effects (Kassenova, 2016). For the people of Semipalatinsk, the cost and stigma of becoming 'radioactive mutants' mean that they have no option but to remain in their contaminated homelands (Stawkowski, 2016). Conversely, the people of Bikini Atoll were forced to permanently relocate to other atolls, and Bikini was left polluted for the foreseeable future by US nuclear weapons (Jorgenson, 2016). The experiences of Australian Aboriginal and Torres Strait Islander peoples, Kiritimati Islanders and Marshallese people offer a representative insight into the diversity of mobility necropolitics that emerged as aspirant nuclear weapon possessor states tested their bombs with casual disregard for Indigenous communities.

The first British atmospheric nuclear test was undertaken on Montebello Island, Australia in 1952. British nuclear testing expanded across South West Australia, to include Maralinga, Emu Field and Montebello Island during the 1950s (Arnold and Smith, 2006). Both British and Australian governments depicted these sites as insignificant wastelands – the flora, fauna and the people who lived there as insignificant under the Australian colonial laws. The Australian Aboriginal and Torres Strait Islander communities who lived there were considered sub-human by the Australian government, and were not counted as citizens until 1967 (Walsh, 1997; Reed and Stillman, 2010; Taylor, 2011). Through this colonial racism, any risk to the people or their lands from nuclear testing was rendered invisible. As there was no explanation, support or guidance by the British military, the Australian Aboriginal and Torres Strait Islander communities had to create their own meanings and narratives to understand the consequences of nuclear weapons.

Maralinga means 'thunder' in the Yolngu language, and these sites had ancient spiritual significance to the Yolngu Nation peoples (Brady, 2017). When they first encountered the wreaths of radioactive fallout that arose from Australian tests, Aboriginal and Torres Strait Islander people described it as 'puyu', the black mist (Eames, 1985; Mittmann, 2017; Williams et al., 2017). For these open-air and previously mobile communities, who remain deeply connected to weather and the environment (see chapter by

Wright et al., this volume), the arrival of puyu was a perturbing phenomenon (Williams et al., 2017).

No explanation was forthcoming from local military encampments to support Aboriginal and Torres Strait Islander communities in protecting themselves from radioactive exposure. As Indigenous people became unwell, the test sites became known as 'Mamu', or evil spirit places. Nyarri Morgan was a local young man at the time of the Maralinga nuclear weapon tests, and he described their impacts to his community (Mann and Morgan, 2016).

> We thought it was the spirit of our gods rising up to speak with us... then we saw the spirit had made all the kangaroos fall down on the ground as a gift to us of easy hunting so we took those kangaroos and we ate them and people were sick and then the spirit left...
>
> The smoke went into our noses, and other people still have that poison today.

Through these Aboriginal and Torres Strait Islander understandings, what was initially interpreted as an act of spiritual weather benevolence had become the powerful retribution of angry gods. As Aboriginal and Torres Strait Islander peoples were not included as Australian citizens, no one knows the true human costs to their communities.

The US began testing nuclear weapons on the Marshall Islands in 1946, despite being tasked with Marshallese people's self-determination after World War II (Beck and Bennett, 2002; Hirshberg, 2012). By sanctioning US control over Micronesian lands, the United Nations (UN) Trusteeship Agreement hastened the rise of an imperialistic nuclear military-industrial complex. When nuclear testing began, hundreds of Marshallese people were forced to leave their homes 'for the good of mankind and to end all wars' (Ruff, 2015). Yet their islands were permanently and uninhabitably contaminated due to radioactive fallout from the US nuclear tests.

Lemyo Abon was displaced from Rongelap Atoll in the Marshall Islands, and described his experience of forced relocation: 'For almost 60 years, we have been displaced from our homeland, like a coconut floating in the sea with no place to call home'. Others were repeatedly relocated by the US military to inappropriate locations such as Rongerik Atoll, a desert island that did not have a lagoon to support their traditional way of life (Parsons and Zaballa, 2017). By 1949, Micronesian and Marshallese labourers were excluded from their own homes on Kwajalein Atoll to make way for picket-fenced military accommodation (Hirshberg, 2012). The US military claimed that suitable living space was no longer available for locals on Kwajalein Atoll and that they must live on Ebeye Island instead. By 1951, the atoll had a singularly American population and the people of Kwajalein Atoll had become foreign commuters to their own lands, entering colonial America by police-escorted ferry each morning and returning to Micronesia each night. Island guides for incoming American employees detailed the availability and cost of Marshallese domestic help, noting that 'They are transported to Kwajalein atoll in the morning and returned to their island at the end of the day' (Hirshberg, 2012). Radioactive waste is still stored on the Marshall Islands in concrete cascades, notably Runit radioactive repository dome on Enewetak Atoll (Connell, 2012; Gerrard, 2015; Guardian, 2016). This repository remains vulnerable to the consequences of climate change (Alexis-Martin, 2019a).

The UK Grapple Megaton Trials began on Kiritimati in 1957, as the British endeavoured to develop a hydrogen bomb to rival the US (Maclellan et al., 2015). The apogee of this test series was Grapple Y, a hydrogen bomb that granted the largest yield of any British thermonuclear device. There was indifference towards the lives of Kiritimati Islanders, who were described as savages. One UK military report stated that

> for civilised populations, assumed to wear boots and clothing, and to wash, the amount of activity to produce this dosage is more than is necessary to give an equivalent dosage to primitive peoples who are assumed not to possess these habits... It is assumed that in the possible regions of fall-out at Grapple there may be scantily-clad people in boats to whom the criteria of primitive peoples should apply.
>
> (Maclellan, 2005)

Taabui Teatata was 11 at the time of the first test. She described how she was unexpectedly moved at midnight by a military commander before the detonation. She was frightened but remembers the army commander who moved her and her family telling her, 'Don't worry, you're safe – this is the British military.' She was loaded onto a ship and taken offshore. She recalls being too frightened to talk: 'It was very crowded, it was meant for cargo and there was no room for children like me to play. There was no space, we were treated like animals.'

These cramped immobilities became the reality of island life during the Cold War. Islanders were corralled within their village, away from military encampments, then moved to ships or tennis courts during the nuclear tests (Alexis-Martin, 2019a). They were instructed to watch Disney movies while onboard, in a gauche attempt to distract them from the sounds and sights of Britain's largest mushroom cloud. This community, whose lives took place under the Pacific sunshine and in harmony with their local environment, was confined and screened from the unnatural weather that was unleashed upon their island.

Nuclear winter

While devastating visions of nuclear war have been a part of our popular culture for decades, there has been a resurgence of cultural interest in nuclear winter (Thompson and Schneider, 1985). We may never see the apocalyptic spectacle of nuclear winter, with the sun dimmed by clouds of nuclear ash and weather systems disrupted across the planet, and we may never have to experience the accompanying (im)mobilities of this scenario; yet, the risk continues to escalate. International nuclear warfare policy is becoming increasingly irregular and multilateral in nature. The Bulletin of the Atomic Scientists' Doomsday Clock was set forward to 100 seconds to midnight on 23rd January 2020, to emphasise this growing risk(Spinazze, 2020). Their statement on the prescience of nuclear threats explored the failure of arms control treaties, and the possibility of a nascent arms race (Bulletin of the Atomic Scientists, 2020).

> In the nuclear realm, national leaders have ended or undermined several major arms control treaties and negotiations during the last year, creating an environment conducive to a renewed nuclear arms race, to the proliferation of nuclear weapons, and to lowered barriers to nuclear war. Political conflicts regarding nuclear programs in Iran and North Korea remain unresolved and are, if anything, worsening. US-Russia cooperation on arms control and disarmament is all but non-existent.

Perhaps nuclear winter is more likely now than ever before. Nuclear winter is the hypothetical outcome of the global environmental and climatic consequences of nuclear war (Turco et al., 1983). These consequences could arise from vast quantities of airborne particulates produced by nuclear warfare and distributed densely enough to dim the sun, cool the earth and induce an unnatural global winter (Turco, 2017). Mutually assured destruction assumes that the full-scale use of nuclear weapons by two or more opposing sides would cause complete annihilation of the protagonist and defendant (Batten, 1966). However, nuclear winter posits that widespread and large-scale nuclear warfare could produce enough pollution to completely change our weather patterns, producing a self-assured destruction (Robock and Toon, 2012). Temperatures could decline dramatically, agriculture could fail, and both human and non-human animals could perish from famine (Scouras, 2019). Our way of life could irrevocably change.

However, this apocalyptic imaginary has been a powerful force for change, notably through the US Senate's 1973 resolution 'to prohibit and prevent, at any place, any environmental or geophysical modification activity as a weapon of war' (Fleming, 2006). This resolution forbade cloud seeding and earthquake induction rather than the climatological conditions arising from nuclear war. It showed an awareness of the immobilising consequences of weather warfare, despite the entanglements of nuclear and weather warfare being not yet realised. Prominent Cold War meteorologist Howard Orville warned that 'If an unfriendly nation gets into a position to control the large-scale weather patterns before we can, the result could even be more disastrous than nuclear warfare' (Fleming, 2006). Interestingly, a science-centred approach obscured the fundamental goal of scientists' nuclear winter campaign: to challenge the politics of the Cold War and end the arms race (Rubinson, 2016). Thus, nuclear winter studies offered gruesome yet internationally peer-reviewed scenarios of cold, dark and fallout – to try to prevent the arms race from reaching a grisly conclusion. This techno-scientifically immobilising action by scientists may have prevented further mushroom clouds from arising, thus retaining Hiroshima and Nagasaki's place as the only sites of nuclear attack to date.

Conclusions

The weather generated by nuclear warfare can shatter and alter the movements and lives of its surrounding communities. The diverse range of human, atmospheric and technological (im)mobilities that arise during nuclear warfare are historically entangled with nuclear imperialism, and representative of the impacts of a nuclear imperialism–necropolitics nexus (Alexis-Martin, 2019a). However, even nuclear winter cannot last forever. The traces of underground nuclear weapon tests are scorched into our stratigraphy, their anthroturbidation cross-cutting deep time, unaffected by the weathering and erosion that blight the evidence of airburst nuclear weapon tests (Zalasiewicz et al., 2014). As such, they will persist millions of years into the future, providing geological evidence of our atomic age, long after the demise of the atomic mushroom clouds and perhaps even after humanity itself.

Notes

1 RDS stands for Rossia Delaet Sama which means 'Russia does it itself'. There was support from German scientists' post-Third Reich and international spies, notably, Ethel and Julius Rosenburg in the UK, so this is not strictly true.
2 Paradise and island life emerged as a theme during my interviews with nuclear test veterans from 2016 to 2018, for my NCCF funded Nuclear Families project.
3 From interviews with nuclear test veterans from 2016 to 2018 for my NCCF funded Nuclear Families project.

References

Adey, P, Bissell, D, Hannam, K, Merriman, P and Sheller, M (eds.) (2014). *The Routledge Handbook of Mobilities*. Abingdon, UK: Routledge.

Alexis-Martin, B (2016). *'It Was a Blast!'—Camp Life on Christmas Island, 1956–1958*. Arcadia.

Alexis-Martin, B (2019a). *Disarming Doomsday: The Human Impact of Nuclear Weapons since Hiroshima*. Radical Geography. London, UK: Pluto Press.

Alexis-Martin B (2019b). The nuclear imperialism-necropolitics nexus: contextualizing Chinese-Uyghur oppression in our nuclear age. *Eurasian Geography and Economics*, 60(2): 1–25.

Alexis-Martin, B and Davies, T (2017). Towards nuclear geography: zones, bodies and communities. *Geography Compass*, 11(9): 1–13.

Aoki, K (1996). Foreign-ness & Asian American identities: yellowface, World War II propaganda, and bifurcated racial stereotypes. *UCLA Asian Pacific American Law Journal*, 4: 1.

Arnold, L and Smith, M (2006). *Britain, Australia and the Bomb: The Nuclear Tests and their Aftermath*. New York: Springer.

Batten, ES (1966). *The Effects of Nuclear War on the Weather and Climate*. Santa Monica, CA: RAND Corporation.

Beck, HL and Bennett, BG (2002). Historical overview of atmospheric nuclear weapons testing and estimates of fallout in the continental United States. *Health Physics*, 82(5): 591–608.

Bernstein, BJ (1998). Truman and the A-bomb: targeting noncombatants, using the bomb, and his defending the 'decision'. *The Journal of Military History*, 62(3): 547.

Brady, M (2017). Atomic thunder: the Maralinga story [Book Review]. *Aboriginal History*, 41: 235.

Connell, J (2012). Population resettlement in the pacific: lessons from a hazardous history? *Australian Geographer*, 43(2): 127–142.

Cresswell, T (2011). Mobilities I: catching up. *Progress in Human Geography*, 35(4): 550–558.

Danielssons, PR (1986). *French Nuclear Colonialism in the Pacific*. Ringwood: Penguin Australia.

Davies, T (2015). Nuclear borders: Informally negotiating the Chernobyl exclusion zone. In *Informal Economies in Post-Socialist Spaces* (pp. 225–244). London: Palgrave Macmillan.

Eames, G (1985). Royal Commission into British Nuclear Tests in Australia: final submission by counsel on behalf of Aboriginal organisations and individuals.

Fleming, JR (2006). The pathological history of weather and climate modification: three cycles of promise and hype. *Historical Studies in the Physical and Biological Sciences*, 37(1): 3–25.

Gerrard, MB (2015). America's forgotten nuclear waste dump in the Pacific. *SAIS Review of International Affairs*, 35(1): 87–97.

Hawkins, HT (2013). *History of the Russian Nuclear Weapon Program* (No. LA-UR-13-28910). Los Alamos, NM: Los Alamos National Lab (LANL).

Hirshberg, L (2012). Nuclear families: (Re) producing 1950s suburban America in the Marshall Islands. *Organization of American Historians Magazine of History*, 26(4): 39–43.

Hoshi, M, Sawada, S, Nagatomo, T, Neyama, Y, Marumoto, K and Kanemaru, T (1992). Meteorological observations at Hiroshima on days with weather similar to that of the atomic bombing. *Health Physics*, 63(6): 656–664.

Ingold, T (2007). Earth, sky, wind, and weather. *Journal of the Royal Anthropological Institute*, 13: S19–S38.

Ingold, T (2010). Footprints through the weather-world: walking, breathing, knowing. *Journal of the Royal Anthropological Institute*, 16: S121–S139.

Jacobs, R (2013). Nuclear conquistadors: military colonialism in nuclear test site selection during the Cold War. *Asian Journal of Peacebuilding*, 1(2): 157–177.

Jones, CA and Smith, MD (2015). War/law/space notes toward a legal geography of war. *Environment and Planning D: Society and Space*, 33(4): 581–591.

Jorgenson, T (2016). Bikini Islanders still deal with fallout of US nuclear tests, 70 years later. *The Conversation*. June 29, 2016. https://theconversation.com/bikini-islanders-still-deal-with-fallout-of-us-nuclear-tests-70-years-later-58567

Kahn, M (2000). Tahiti intertwined: ancestral land, tourist postcard, and nuclear test site. *American Anthropologist*, 102(1): 7–26.

Kant, I (1987). *Critique of Judgment 1790*. Trans. Werner S. Pluhar. Indianapolis: Hackett.

Kassenova, T (2016). Banning nuclear testing: lessons from the Semipalatinsk nuclear testing site. *The Nonproliferation Review*, 23(3–4): 329–344.

Keown, M, Taylor, A and Treagus, M (eds.) (2018). *Anglo-American Imperialism and the Pacific: Discourses of Encounter*. Abingdon, UK: Routledge.

Liebow, AA (1983). Encounter with disaster: a medical diary of Hiroshima, 1945. Condensed from the original publication, 1965. *The Yale Journal of Biology and Medicine*, 56(1): 23.

Mbembe, A (2019). *Necropolitics*. Durham, NC: Duke University Press.

Maclellan, N (2005). The nuclear age in the Pacific islands. *The Contemporary Pacific*, 17(2): 363–372.

Maclellan, N, Deery, P and Kimber, J (2015, February). Grappling with the Bomb: opposition to Pacific nuclear testing in the 1950s. In Proceedings of the 14th Biennial Labour History Conference (p. 21). Australian Society for the Study of Labour History.

Mann, A and Morgan, N (2016). Aboriginal man's story of Maralinga nuclear bomb survival told with virtual reality. *ABC News Media*. Australia. Accessed 11th November 2019. https://www.abc.net.au/news/2016-10-07/aboriginal-mans-story-of-nuclear-bomb-survival-told-in-vr/7913874

Masco, J (2008). Nuclear techno-aesthetics: sensory politics from trinity to the virtual bomb in Los Alamos. *American Ethnologist*, 31(3): 349–373.

Massey, D (1993). Power-geometry and a progressive sense of place. In *Mapping the Futures: Local Cultures, Global Change* (pp. 59–69). Abingdon, UK: Routledge.

Matsunari, Y and Yoshimoto, N (2013). Comparison of rescue and relief activities within 72 hours of the atomic bombings in Hiroshima and Nagasaki. *Prehospital and Disaster Medicine*, 28(6): 536–542.

Merleau-Ponty, M (1963). *The Structure of Behavior*. Boston: Beacon Press.

Michishita, T (2017). After the bomb: survivors of the atomic blasts in Hiroshima and Nagasaki share their stories. *Time*, 5th August 2017. Accessed on 20th December 2019.

Mittmann, JD (2017). Maralinga: aboriginal poison country. *Agora*, 52(3): 25.

Nishii, M (2019). Out of the destruction of Hiroshima: the social history from primary sources of rebuilding human lives during the city's reconstruction. *Journal of the Asia-Japan Research Institute of Ritsumeikan University*, 1: 16–28.

Nye, David E (1994). *American Technological Sublime*. Cambridge, MA: MIT Press

Parsons, KM and Zaballa, RA (2017). *Bombing the Marshall Islands: A Cold War Tragedy*. Cambridge, UK: Cambridge University Press.

Reed, TC and Stillman, DB (2010). *The Nuclear Express: A Political History of the Bomb and Its Proliferation*. London, UK: Zenith Press.

Robock, A and Toon, OB (2012). Self-assured destruction: the climate impacts of nuclear war. *Bulletin of the Atomic Scientists*, 68(5): 66–74.

Rosenthal, P (1991). The nuclear mushroom cloud as cultural image. *American Literary History*, 3(1): 63–92.

Roy, A (1998). *The End of Imagination*. New York: DC Books.

Rubinson, P (2016). Imagining the apocalypse: nuclear winter in science and the world. In *Understanding the Imaginary War*. Machester, UK: Manchester University Press.

Ruff, TA (2015). The humanitarian impact and implications of nuclear test explosions in the Pacific region. *International Review of the Red Cross*, 97(899): 775–813.

Rush-Cooper, N (2013). *Exposures: Exploring selves and landscapes in the Chernobyl Exclusion Zone.* Doctoral dissertation, Durham University.

Russell, C (1975). The weather as a secret weapon: from Vietnam to Geneva. *Washington Star*, August 23, 1975, reprinted in Senate committee on foreign relations, Subcommittee on oceans and international environment, Prohibiting military weather modification: hearings on S.R. 281, 92nd Cong., 2nd sess., 1972, 47.

Schäfer, S (2016). From Geisha girls to the atomic bomb dome: dark tourism and the formation of Hiroshima memory. *Tourist Studies*, 16(4): 351–366.

Scouras, J (2019). Nuclear war as a global catastrophic risk. *Journal of Benefit-cost Analysis*, 10(2): 274–295.

Spinazze, G (2020). https://thebulletin.org/2020/01/press-release-it-is-now-100-seconds-to-midnight/ Accessed 23rd January 2020.

Stawkowski, ME (2016). 'I am a radioactive mutant': emergent biological subjectivities at Kazakhstan's Semipalatinsk Nuclear Test Site. *American Ethnologist*, 43(1): 144–157.

Taylor, J (2011). Postcolonial transformation of the Australian Indigenous population. *Geographical Research*, 49(3): 286–300.

The Guardian (2016). Marshall Islands nuclear arms lawsuit thrown out by UN's top court. 6th October 2016. https://www.theguardian.com/world/2016/oct/06/marshall-islands-nuclear-arms-lawsuit-thrown-out-by-uns-top-court

Turco, R (2017). Nuclear foreboding: shadows cast by nuclear winter. *Bulletin of the Atomic Scientists*, 73(4): 240–243.

Thompson, SL and Schneider, SH (1985). Nuclear winter reappraised. *Foreign Affairs*, 64: 981.

Turco, RP, Toon, OB, Ackerman, TP, Pollack, JB and Sagan, C (1983). Nuclear winter: global consequences of multple nuclear explosions. *Science*, 222(4630): 1283–1292.

Varga, HH (2017). Arundhati Roy, the end of imagination. *Romanian Journal of Indian Studies*, (1): 125–132.

Walsh, J (1997). Surprise down under: the secret history of Australia's nuclear ambitions. *The Non-proliferation Review*, 5(1): 1–20.

Washington Post (1945) War Department Press release on the day of the Hiroshima bombing. Brig. General Thomas F. Farrell's description of the 16th July atomic bomb test in New Mexico, 7th August 1945.

Wellerstein, A (August 7, 2015.). Nagasaki: the last bomb. *The New Yorker*.

Williams, GA, O'Brien, RS, Grzechnik, M and Wise, KN (2017). Estimates of radiation doses to the skin for people camped at Wallatinna during the UK Totem 1 Atomic Weapons Test. *Radiation Protection Dosimetry*, 174(3): 322–336.

Yui, Y (2003). Changing characteristics of public housing dwellers in Hiroshima City. *Geographical Review of Japan*, 76(5): 333–348.

Zalasiewicz, J, Waters, CN and Williams, M (2014). Human bioturbation, and the subterranean landscape of the Anthropocene. *Anthropocene*, 6: 3–9.

Zwigenberg, R (2016). The atomic city: military tourism and urban identity in postwar Hiroshima. *American Quarterly*, 68(3): 617–642.

17 Writing (extra)planetary geographies of weather-worlds

Kimberley Peters

In the beginning: introducing extra-planetary relations

If you listen to the recording of the sounds of winds from the planet Mars, on a video offered by NASA (the National Aeronautics and Space Administration of the United States), you would be forgiven for thinking you were just listening to a sound recording of a fairly blowy day on planet Earth. There is something pedestrian, yet spectacular, about the sound. It is the sound we expect of wind when it whistles in the ears: the sound of air whipping up and whooshing past. Sound, as myself and Geraint Whittaker have noted, 'occurs when something, anything, vibrates. For there to be sound, there must be vibration' (in-press). What we may hear, on a day-to-day basis, is the result of vibrations – tiny movements (Bissell 2010) – which translate into something which can be heard (via the cochlea) when travelling (most often) through the medium *of* air (Sharp 2017). Consider 'all the vibrations which become sound: the drop of a plate, a shout in the street, the rev of an engine, a cough. Sound doesn't hang around long (even in an echo)' (Whittaker and Peters, 2021). But when we hear the wind, it is the movement, or mobility of air *itself* amidst the spaces and places it touches, passes over, and rattles through, which creates that familiar 'windy' noise that becomes recognisable to us. The sound we know as wind, becomes recognisable to us through the repeated movements, or the *mobilities*, of air through its surrounds (Atkinson 2007). We become attuned to the whoosh and whistle 'of the wind' so that when we hear it, even when sitting comfortably inside, we might confidently say, 'it *sounds* windy outside'.

The wind on Mars has that same-sounding whoosh and whistle of the movement of Earthly air. Yet in December 2018, when NASA's InSight probe – which successfully landed on the 'Red Planet' on 26th November of the same year – accidentally started to pick up sounds of an extra-planetary wind, it was also rather remarkable. Humans were listening to another planet, and its weather-worlds, for the first time. What was more remarkable was that the sounds were heard via the seismometer, the technological apparatus, which was intended to make readings of *ground movements*, not aerial ones. It was designed to measure Marsquakes not airquakes. The whipping of the wind around the instrument caused small vibrations, translating the sound of that distant wind, back to Earth. It was both extraordinary in its unexpected capture by NASA recording devices, and in spite of its recognisably 'windy character', it was special in being an altogether previously *unheard* sound. We, on planet Earth, heard extra-terrestrial weather through the albeit second-hand, recycled recordings of Martian vibrations through the air. We were listening to distant weather: weather from another space, another time, and another *world*.

Taking-off from the example of the InSight probe, this chapter acts as a reminder that weather is not confined to our own planetary limits. Going 'off-Earth' allows us to probe other insights into the nexus between weather, mobilities and space (in this case, quite literally *outer space*). As scholars have contended (see Cosgrove 1994; Gilroy 2003; Jazeel 2011), the planetary and extra-planetary view (sometimes problematically) offers us a different kind of perspective to worldly life. Indeed, this vantage point both ties together humanity in reflecting the state of the world as a whole, planetary unit. Yet this scale also obscures differences, divergences and discriminations that shape our planet with such a homogenising perspective. Although this chapter stops short of exploring some of these lived tensions of the (extra)planetary, it is necessary to acknowledge how discourses of the planetary, Earthly and extra-terrestrial are configured – and that conceiving of weather through this frame is not an objective, scientific enterprise, but one that encompasses people, places and politics.

What this chapter does focus on, through the vantage of (outer) space, are the broader relationalities bundled up in our weather-worlds, *moving* our notions of weather into deeper times and spaces. In doing so it redresses the typical notions of proximity, distance, temporality, speed and scales of weather and its mobilities. To do so, in what follows, this chapter tracks through the possibilities of writing new (extra)planetary geographies of weather-worlds. The chapter begins by providing a brief overview of work in geography and the wider social sciences which is turning to 'space' as a site of research interest. It further connects this ever-burgeoning field of work with research that has attempted to link mobilities and off-Earth worlds; building the case for examinations of weather mobilities that stretch beyond (yet back to) Earth. It will then cover two ways of thinking about weather, mobilities, outer space and Earth. First it will move to *deep space*, considering scale and how weather on planet Earth can be thought of anew through solar system relations. Second it will move to *deep time*, to examine how weather on planet Earth is determined by the temporalities of a broader solar system and universe in motion, including the motion of Earth itself. Finally, the chapter concludes by moving to focus on weather mobilities 'beyond' Earth and the distant planetary weathers of Venus and Jupiter to consider what the (extra)planetary perspective may bring (up) that confirms and confronts our ways of knowing how to live in the world, amidst weather-extremes, climate change and irreparable Anthropocenic environmental harm.

Take me to the moon (and beyond): going into orbit

Outer space, the universe and extra-planetary and terrestrial worlds are not only the remit of natural sciences. There has been a keen 'turn' towards understanding our wider relations with the solar system, distant galaxies, and the 'beyond', in the social sciences and humanities (Beery 2016; Dixon 2018; Dunnett et al. 2019; Jones 2011; Kitchin and Kneale 2005; MacDonald 2007; Peters et al. 2018). In geography, the study of outer space has long been part of the discipline's 'early modern heritage', with the work of Elijah Burritt, for example, mapping the 'heavens' and celestial realm in the late 19th century (see MacDonald 2007: 595), and later, the theories of early 20th-century scholars such as Mackinder (with his 'pivot' and 'heartland' concepts) being adopted to make sense of extra-planetary 'space race' relations (see Dolman 2005). Yet, in spite of early beginnings, MacDonald has argued of a lack of sustained scholarship concerned with the cosmos in contemporary social sciences, the humanities and human geographies (2007) and, more recently, Dunnett et al. note the need for more engagements with space as it becomes enveloped 'with people on earth in various social, cultural and economic contexts' (2019: 317).

There have been moves to thinking about engagements with space through the framework of mobilities scholarship in recent years, unpacking the 'politics of movement' (Cresswell 2006, 2010; Sheller and Urry 2006) in relation to the possibility of extra-terrestrial travels for traditionally Earth-bound citizens. Cohen, for example, explores the futures of space mobilities for humans, considering new proposed forms of off-Earth tourism. Here he examines the discourses of 'adventuring' that underpin (colonial) expansions and exploitations of space by humans (2017). Similarly, Damjanov and Crouch explore the mobilities of people to space via virtual means, allowing access to distant 'planetary exteriors' in ways that challenge our 'collective ways of traveling and seeing, performing and consuming and configuring our mediated and embodied senses of place' (2017: 1). In a different frame, Spector et al. (2017) focus on space flight to confront issues around planetary sustainability, exploring the tensions that exist between 'modern mobilities' and the 'biosphere' (the worldwide scale of Earth where all life exists), which may emerge when we bring space travel into play.

In spite of a burgeoning of extra-planetary work (see Dunnett et al. 2019 for a review) and research that is beginning to draw lines between space and mobilities (Cohen 2017; Spector et al. 2017), within this turn, there has not – to date – been a discussion of how weather, mobilities and space coalesce. Such a discussion is arguably pertinent to make sense of how weather, as a mobile condition, stretches or reaches beyond our immediate surrounds, linking to wider, deeper processes and space-times which shape both the mobilities of weather, and our mobilities in relation. Indeed, *moving* studies of weather mobilities beyond Earth, yet still thinking relationally towards our Earth, allow us to consider how weather moves, is mobile, and relates to mobilities in previously unexamined ways that may bring to light new ways of thinking, knowing and understanding such phenomenon. It is not only the social, cultural, and economic relations 'with people on earth' which we must explore through space, it is also our geophysical, weather-relations.

Here comes the sun: mobilities in deep space

Weather, as we know from elsewhere in this book, is concerned with the 'mix of events that happen each day' *inside* Earth's atmosphere (UCAR 2020). Weather, then, has a spatiality *within* the geographic, territorial limits of Earth. It is a phenomenon that is typically understood as *contained* in our planetary sphere and relates to a set of physical, mobile processes linked to Earth's elements: water and air, particularly the level of the troposphere (Adams-Hutcheson 2019). The troposphere is the lowest atmospheric strata or layer of Earth. It is here where changes in air pressure and water vapour contents result in (fairly) frequently changing conditions of temperature, wind speed, precipitation and humidity. It is a site and space of mobility, where movements of air and water vapour molecules – movements of invisible matter (Anderson and Wylie 2009) – alter, blend, rotate to create weather. Indeed, the word 'troposphere' comes from the Greek 'tropein – to change, circulate or mix', and it is the space where 'most of the weather phenomena, systems, convection, turbulence and clouds occur' (Weather Online 2020).

Because of the conditions of 'change, circulation and mixing' in the troposphere, weather can differ spatially (sometimes radically), at any given moment, but often (mostly relevantly to us) at lower levels of the scalar hierarchy, the regional or local scale: it may be sunny across the north-west of England but wet in London, or raining in my street but dry in the next. The capacities of air and water vapour to act at these scales lead us to think of weather *within* these spatial parameters, *within* Earth. Weather forecasts tend to give us regional predictions of general weather patterns, even though specific localised conditions may differ slightly, and weather experiences come about via localised embodied engagements that confirm or contest these predictions through our entanglements with the very matter of weather (Ingold 2008). Literature on our connections with weather, then, seem to place its mobilities and our mobilities in and through it within a spatial frame of the local, or the space 'closest in' – the body (see Adams-Hutcheson 2019; Barry 2019; Martin 2011; Peters 2012; Simpson 2019; Larsen and Jenson, *this collection*).

But thinking of weather at the scale of the troposphere takes us beyond the more regionalised, localised or embodied experiences of weather, where weather is often a condition 'felt', 'lived' and 'walked-through' via a mixing of person–ground–earth–air–sky, as Tim Ingold puts it (2010). It takes us beyond, to the broader tropospheric – planetary – scale, where the troposphere is an atmospheric layer stretching across the globe. This is important for rethinking our relations with weather. As Adams-Hutcheson notes, the everyday experiences of farmers in Aotearoa New Zealand and their attachments and engagements with weather are tropospheric entanglements (2019). Although Adams-Hutcheson focuses at the *lived* experiences of daily life around the practice of farming, she does so within a broader conceptual scalar knowledge of weather-worlds (Ibid.). Turning to Ingold here is important for (re)thinking the scale of weather on planet Earth. Ingold's perspective is, on the one hand, like many weather scholars, a perspective from the intimate, embodied, felt scale of human experience (2010). On the other hand, his conceptualisation of weather (and our place within in) is through an understanding of being 'in the open' (2007). For Ingold, our entanglements with the weather-world eradicate the 'line' between Earth and sky, ground and air, because through the acts of breathing, walking, being, we are enveloped within an 'open' world where we mix and mingle with the elements around us (2007: 19).

But does Ingold go far enough? His is a predominantly *Earth-bound* notion of 'openness'. Whilst Ingold cautions against boundaries, borders and the separations of the worldly spheres such as Earth, sky and sea, he focuses (surprisingly, and perhaps ironically) on the openness of weather-worlds *inside* or *within* our planetary limits. In this way we can understand weather not so much as an external force upon us, or a set of conditions 'originating' from the sky to the ground, but part of an open world of mixing and mingling. As Ingold writes, '[t]he open world... has no... boundaries, no insides or outsides, only comings and goings' (2008: 1801). Yet Ingold's 'open world' is oddly closed to outer space.

Building from Ingold – and expanding his viewpoint further – we may understand that we are not only entangled in planetary 'becomings' but *extra-planetary* ones. Ingold's ideas of openness allow us, I argue, to more radically consider an extra-planetary openness and mingling with *deep space*, through our experiences of the weather-world. In Ingold's 'weather-world', his *Earthly* world, 'there is no distinct surface separating earth and sky. Life is rather lived in a zone in which substance and medium are brought together in the constitution of beings which, in their activity, participate in weaving' (2008: 1804). However, it could be put that there is also no distinct surface separating Earth and (outer) space. Although, like Earth and sky, the spaces appear geophysically different and distinct, Ingold shows how they are bound together. Likewise, Earth – in this imaginary, and ontology – is bound or woven together with the wider universe in the openness of space.

Such a way of thinking *openly* takes us (literally) into deep space, bringing us into touch with unimaginable scales of extra-planetary proportions, which shape our experiences and entanglements with weather. Indeed, all weather is driven by movements of radiation from the sun through the solar system. All weather, as Lewis notes, 'is solar-powered' (Lewis 2020; see also Vannini and Taggart 2015). As Lewis goes on to note, our star, the sun, is the heat source which is bound together with the atmosphere within our planet. He states,

> The amount of power radiated by the Sun is enormous, roughly 1.4 kW/ m2 at the distance of the Earth. Allowing for scattering back to space by clouds and ice, and on average over the year and time of day, about 200 W/m2 of this solar radiation reaches the Earth's surface.
>
> (Lewis 2020: n.p)

It is this huge source of heat which entangles with the air to change pressure and to create water vapour. Without the sun, there is no weather. As such, when we make a statement like 'here comes the sun' – referring to the presence of locally, bright, warm weather and of the feel of heat on our skin – the sun is not so much 'coming' but *becoming,* with us, as part of an openness that is spun from a network which is broader than we typically account for. Our regional, local, embodied weather-worlds are thus also extra-terrestrial weather-worlds. Perhaps then, to follow Ingold (2007, 2008), the idea of scale is less useful – rather weather should be thought of as a condition of openness, and of *extra-planetary* openness. Weather comes (in) to be(ing) via the mobilities of rays not *external* to the planet (conceived as the 'outside' coming 'inside') but as a 'sticky web' of bindings between our planet and a wider network and assemblages of deep spaces, sites and energies.

Although tropospheric weather is churning and changing, the sun alerts us to something of a constant in our weather-world – a sometimes seemingly absent, but always present part of our relations with weather. The star is always burning, the sun is always shining. Indeed, 'the Sun is this huge mass of energy and it's *constantly* emitting that energy out into space' (New World Climate 2020). Rays come constantly, repeatedly, over time, to our Earth, as part of the making of our weather-systems. Being ourselves more open to the openness of our weather-worlds is important for understanding how we think about the weather 'worlds' we experience. It also helps us to reconceptualise the temporalities of weather – where weather seems to be a constantly changing set of conditions. As such, although Massey has noted that '[t]he most predictable remark you can make about the weather concerns, of course, its unpredictability' (Massey 2003: n.p), perhaps, when we think of weather's *deep space* relations, the most unpredictable remark we can make about the weather concerns, surprisingly, its *predictability.*

I'm so dizzy: mobilities in deep time

Turning to temporalities alerts us to weather relations that happen not just in deep space but deep *time.* Deep time is a term linked to the writer John McPhee and the book *Basin and Range* (1982), where he delves in the worlds of geology and geological timescales that are so 'deep', they are beyond human comprehension (although the term is said to have longer, deeper roots to the 18th-century Scottish scientist James Hutton, see Farrier and Aeon 2016). Deep time is time that has a far distant reach – into the past, but also into the future. As Farrier and Aeon note,

> Hutton posited that geological features were shaped by cycles of sedimentation and erosion, a process of lifting up then grinding down rocks that required timescales much grander than those of prevailing Biblical narratives. This *dizzying* Copernican shift threw both God and man into question. 'The mind seemed to grow giddy by looking so far back into the abyss of time,' was how John Playfair, a scientist who accompanied Hutton on several crucial expeditions, described the effect of looking over the stratified promontory of Siccar Point in Scotland.
>
> (Farrier and Aeon 2016, emphasis added)

The idea of time so distant and far into the past was to make scientists 'dizzy' along with the (by this point) understanding of a different mode of *dizzying,* orbital, temporal relations, which were said to form and forge planetary life. The *Copernican* model (named after the scientist Nicolaus Copernicus) proposed a temporal reality based on Earth's movement around the sun (and the movement of other planets around the sun too, in a heliocentric model). The model also proposed the movement of Earth itself on its axis. It was a radical shift in thinking in the sixteenth century, away from the classical ontology of a static Earth in relation to the heavens, where other celestial bodies travelled through a zodiacal course around Earth (see Rabin 2004).

The circular movement of Earth in orbit, and Earth itself as spinning, all in relation to the sun as the centre of our solar system, has had deep time changes on the conditions of our planet. The orbit of Earth in open relationality to the sun (and other planets, namely the gravity pull of Jupiter) is fairly but not completely constant. Earth's orbit changes slightly over deep time frames from a 'nearly circular to elliptical' shape – what is known as 'eccentricity' (Maslin 2016: 208). On a (very) simple level, eccentricity brings Earth, at times, closer to the sun. As Maslin outlines, knowledge of eccentricity (developed by Milutin Milanković, in what were to become known as Milankovitch cycles) was researched by others to shine light (literally) on how orbital forms, over many thousands of years, related to ice ages (Hays et al. 1976). Moreover, one of Milanković's other cycles, 'obliquity', refers to the (variable) 'tilt of Earth's axis of rotation with respect to the plane of its orbit' again, impacting the closeness of some regions of Earth to the sun over deep time frames (see Maslin 2016).

It might be asked what these deep time, circular, elliptical, tilting, Earthly movements have to do with weather (they seem more relevant to defining planetary 'ages' and alterations in climate – the long-term patterning of weather). But thinking of such deep times and the 'dizzying' movements of our Earth in orbit and its rotation on its own axis at variable tilts, alerts us to deeper ways to think about weather relations. Although these mobilities, which are far more complex than described here (see Maslin for greater detail, 2016), are on a temporal frame that seems irrelevant to the churning, change of weather around us, the movement of Earth in space, is deeply connected to our everyday planetary weather-worlds. As Buis notes,

> Our lives literally revolve around cycles: series of events that are repeated regularly in the same order. There are hundreds of different types of cycles in our world and in the universe...(these) cycles... play key roles in Earth's short-term weather and long-term climate.
>
> (Buis 2020)

For Buis, a NASA scientist, the temporal cycles of our lives – the regular, repeated patterns that we claim as climate, are also part of 'short-term weather' (2020). This is because at (deep) times when Earth is rotationally closer to the sun, there is (to return to the previous section of this chapter) greater heat, or radiation, from the sun entering our planet, shaping atmospheric conditions – *shaping weather.* At times when Earth – and our place in it – is further away, our immediate atmospheric conditions, again, alter. We might think of weather, in deep time, as attuned to the 'rhythms' (Edensor 2010, 2012; Jones 2011) of the universe. As such, wide time frames and deep time have localised, immediate, everyday bearings. Moving beyond Earth, yet still thinking relationally towards it, allows us to consider weather as part of that wider web of openness (Ingold 2008), not just spatially but *temporally* in previously unexamined ways.

Deep time thus allows us to think of the *multiple* mobilities of weather and its (ongoing) formation: adding non-linear and overlapping temporalities to our understandings of weather and our engagements with it. The weather is both 'here' in this immediate time, but simultaneously placed 'there', in another time, *a deep-time*, of Earthly, extra-planetary mobilities. Here Massey's ideas of multiplicity are useful. For Massey, in her analysis of space and time (and space-time – for she does not conceive as either separate or discrete), she moves us away from singular conceptions: away from closed, bounded notions of space, and linear, and timeless 'takes' on time (see Massey 1999, 2001). Conditions of life take place in multiple space-times. For Massey (like Ingold) there is an openness, where such openness alerts us to multiplicity. Weather mobilities are mobilities that exist in manifold space-times (Massey 1999). As Massey herself has noted,

> Each day the newspaper I take provides an account of weather conditions 'Around the world' on the previous day (except in the Americas, because the newspaper is put to bed before the Earth has turned far enough upon its axis to enable noon-time readings there to be registered, read and reported on. For those parts of the world the reading is for the day before yesterday). 88° F and sunny in Algiers; 48° and cloudy in Helsinki; cloudy again but 77° in Beijing. (That day in London it was

57°, with thunder). The coeval existence of a multiplicity of conditions: that is the gift of space. Space is the sphere of the possibility of the existence of plurality, of the co-existence of difference. It is the sphere of the possibility of the existence of more-than-one. Without space there is no 'multiplicity'.

(Massey 2003)

If we follow Massey (1999, 2001) then we can understand weather mobilities as emergent, ongoing, spatially multiple phenomena. Yet these mobilities are also space-*time* phenomena. The multiple conditions of weather played out spatially, are also multiple temporally, *as deep-time and near-time collide.* I thus argue, along with Massey (and through a reading of Massey), that multiplicity is key to our understandings of the mobilities of weather-worlds. Yet, I depart from Massey in shifting part of our understanding of this multiplicity *off-Earth.*

When we move Massey's ideas about process, openness and multiplicity into orbit and when we think through deep time (or rather deep space-time), that which appears stable and unchanging, *moves* (see also Massey 2005). Orbits, which appear regular and consistent over deep time frames, through space, relate to constantly changing atmospheric conditions resulting in a *multiplicity of weather.* In the first part of this chapter, I argued that Ingold's openness might be opened further to encapsulate outer space in this understanding of a 'world' (or rather universe) of bindings. In this section, I have built from Massey to posit that thinking of an Earth of openness and multiplicity – webbed together with the solar system and wider universe – allows us to reconfigure how we think about the relations between mobilities and weather.

2,000 light years from home: mobilities off-earth

At the start (and at the close) of Andy Weir's popular, best-selling book *The Martian* (2014), the protagonist Mark Watney experiences weather extremes whilst stranded on Planet Mars, existing over hundreds of Sols (or Mars days) in a mission to get home. During his solo journey across the 'red planet' he moves (in his Rover) around, and in relation to Mars' winds that whip up red planetary matter in vicious dust storms, which cloud his visibility and threaten to damage his equipment. It is the weather on Mars that leads to his predicament in the first place (where the crew depart without him in a severe storm), and it is weather that almost thwarts his return as he attempts to travel around a mobile, churning storm on his way to the 'MAV' to fly into orbit, and back 'home' to Earth. At the end of the book, the character attempts to track the movement of the storm, to circumnavigate its breadth, its material qualities and mobilities, in a bid to return to Earth.

Weir's book, and his vivid descriptions of weather on Mars, whilst fictional, are a reminder that weather is not limited to our planet. Weather occurs on all planets and in space itself (so-called solar weather). Weather is impacted not just by the gaseous (or lack of gaseous) composition of the atmosphere (Mercury, for example, has no atmosphere) but also by gravity and planetary-level rotation (Lewis 2020). Scientists have sought to study and understand the weather of other planets as a way of better understanding (relationally) our place in the solar system and universe and the conditions through which life is possible (Lewis 2020). NASA, for example, note how Mercury's place in the solar system results in scorching hot weather; how the atmosphere of Venus creates large clouds that trap heat; how Mars cannot hold heat leading to unimaginably low temperatures at night, and Jupiter – the planet of gas – is home to long-raging and moving weather systems (not least evident in Jupiter's 'Great Red Spot', a storm thought to have been in existence for over 300 years) (NASA 2020).

Off-Earth weather reminds us we exist not just in a planetary vacuum but amidst wider, universal relations which also have a bearing on our weather and entanglements with it. This chapter has aimed to think about weather through planetary and extra-planetary lenses, to write different kinds of geographies of 'our' weather-worlds. It has certainly simplified the science where deeper engagements with Earth's movements on its axis and in orbit, and of movements of energy *to* Earth, and relations of gravity *with* Earth could reveal yet more of weather mobilities *on* Earth. However, it has – in the spirit of openness and multiplicity – opened the door to thinking differently about our weather-world(s). Yet, it is important to also make a note of caution about thinking with and through the lens of the 'planetary' and 'extra-planetary'. Although posited here as modes of thinking differently, these terms, frames and ontologies can work to assume a universality in experience. As Jazeel notes, the planetary is often used as a means to bring humanity – and space – together under one remit, or goal (2011). This universalising tendency (a *cosmo*politanism) both works to gloss over difference (see Yusoff on the universal 'we', 2018) as well as veer away from the politics of the planetary – a worldly imaginary that is actually rooted in European, Westernised knowledges and 'ways of seeing' (Jazeel 2011; Cosgrove 1994). The planetary (and extra-planetary) have become important terms in Anthropocene thinking for uniting people around environmental concerns – climate changes and extreme weathers – but they should be deployed critically and carefully. Indeed, the planetary and extra-planetary weather mobilities of this chapter – the deep space and time movements of solar radiation and Earthly tilts – are not felt universally, by individuals, communities, societies. Radiation from the sun is felt more where ozone depletion and its knock-on effects on atmosphere have been most influential or where our increase in gases that keep heat inside the planet have led to weather extremes that have (already) impacted the most vulnerable. To go full circle – or full orbit – then, the (extra)planetary perspective is necessary in understanding how we may reinforce but also challenge our ways of knowing how to live in the world, a mobile world, of irreparable change.

Acknowledgements

Many thanks to Dr Gareth Hoskins for fuelling my interest in the geographies of outer space. This chapter would not have been possible without our collaborations on the cosmos. This work was funded by HIFMB, a collaboration between the Alfred-Wegener-Institute, Helmholtz-Center for Polar and Marine Research, and the Carl-von-Ossietzky University Oldenburg, initially funded by the Ministry for Science and Culture of Lower Saxony and the Volkswagen Foundation through the "Niedersächsisches Vorab" grant program (grant number ZN3285).

References

Adams-Hutcheson, G (2019). Farming in the troposphere: drawing together affective atmospheres and elemental geographies. *Social and Cultural Geography*, 20(7), 1004–1023.

Anderson, B and Wylie, J (2009). On geography and materiality. *Environment and Planning A*, 41(2), 318–335.

Atkinson, R (2007). Ecology of sound: the sonic order of urban space. *Urban Studies*, 44(10), 1905–1917.

Barry, K (2019). More-than-human entanglements of walking on a pedestrian bridge. *Geoforum*, 106, 370–377.

Beery, J (2016) Unearthing global natures: outer space and scalar politics. *Political Geography*, 55, 92–101.

Bissell, D (2010). Vibrating materialities: mobility–body–technology relations. *Area*, 42(4), 479–486.

Buis, A (2020). Milankovitch (orbital) cycles and their role in earth's climate. Available at: https://climate.nasa.gov/news/2948/milankovitch-orbital-cycles-and-their-role-in-earths-climate/ (accessed 03 March 2020).

Cohen, E (2017). The paradoxes of space tourism. *Tourism Recreation Research*, 42(1), 22–31.

Cosgrove, D (1994). Contested global visions: one-world, whole-earth, and the Apollo space photographs. *Annals of the Association of American Geographers*, 84(2), 270–294.

Cresswell, T (2006). *On the Move: Mobility in the Modern Western World*. London: Routledge.

Cresswell, T (2010). Towards a politics of mobility. *Environment and Planning D: Society and Space*, 28(1), 17–31.

Damjanov, K and Crouch, D (2018). Extra-planetary mobilities and the media prospects of virtual space tourism. *Mobilities,* 13(1), 1–13.

Dixon, D (2018). Repurposing feminist geopolitics: on estrangement, exhaustion and the end of the solar system. *Dialogues in Human Geography*, 8(1), 88–90.

Dolman, EC (2005). *Astropolitik: Classical Geopolitics in the Space Age*. London and New York: Routledge.

Dunnett, O, Maclaren, AS, Klinger, J, Lane, KMD, and Sage, D (2019). Geographies of outer space: progress and new opportunities. *Progress in Human Geography*, 43(2), 314–336.

Edensor, T (2010). Walking in rhythms: place, regulation, style and the flow of experience. *Visual Studies*, 25(1), 69–79.

Edensor, T (ed.) (2012) *Geographies of Rhythm: Nature, Place, Mobilities and Bodies.* Ashgate: Farnham.

Farrier, D and Aeon. (2016). How the concept of deep time is changing. *The Atlantic,* Available at: https://www.theatlantic.com/science/archive/2016/10/aeon-deep-time/505922/ (Accessed 03 February 2020)

Gilroy, P (2003). Where ignorant armies clash by night: homogeneous community and the planetary aspect. *International Journal of Cultural Studies,* 6(3), 261–276.

Hays, JD, Imbrie, J and Shackleton, NJ (1976). Variations in the Earth's orbit: pacemaker of the ice ages, *Science,* 194, 1121–1132.

Ingold, T (2007). Earth, sky, wind, and weather. *Journal of the Royal Anthropological Institute,* 13, 19–38.

Ingold, T (2008). Bindings against boundaries: entanglements of life in an open world. *Environment and Planning A,* 40(8), 1796–1810.

Ingold, T (2010). Footprints through the weather-world: walking, breathing, knowing. *Journal of the Royal Anthropological Institute,* 16, 121–139.

Jazeel, T (2011). Spatializing difference beyond cosmopolitanism: rethinking planetary futures. *Theory, Culture and Society,* 28(5), 75–97.

Jones, O (2011). Lunar–solar rhythmpatterns: towards the material cultures of tides. *Environment and Planning A,* 43(10), 2285–2303.

Kitchin, R and Kneale, J (eds.) (2005) *Lost in Space: Geographies of Science Fiction.* London: Continuum.

Larsen, J and Jensen, OB (2020). Marathon running in the 'weather'. In Barry, K, Borovnik, M and Edensor, T (eds.), *Weather Mobilities.* London: Routledge.

Lewis, S (2020). The Weather on all 8 planets of our solar system. From Open Learn at the OU, Available at: https://www.open.edu/openlearn/science-maths-technology/astronomy/the-weather-on-all-8-planets-our-solar-system (accessed 20 April 2020)

MacDonald, F (2007). Anti-Astropolitik—outer space and the orbit of geography. *Progress in Human Geography,* 31(5), 592–615

Martin, C (2011). Fog-bound: aerial space and the elemental entanglements of body-with-world. *Environment and Planning D: Society and Space,* 29(3), 454–468.

Maslin, M (2016). Forty years of linking orbits to ice ages, *Nature,* 540, 208–210.

Massey, D (1999). Space-time, 'science' and the relationship between physical geography and human geography. *Transactions of the Institute of British Geographers,* 24(3), 261–276.

Massey, D (2001). Talking of space-time. *Transactions of the institute of British Geographers,* 26(2), 257–261.

Massey, D (2003) Some times of space. In *Olafur Eliasson: The Weather Project.* Edited by Susan May. Exhibition catalogue. London: Tate Publishing. Available at: http://www.f-i-e-l-d.co.uk/writings-violence_files/Some_times_of_space.pdf (Accessed 03 February 2020)

Massey, D (2005) *For Space.* London: Sage.

NASA. (2018). Sounds of Mars: NASA's InSight senses Martian wind. Available at: https://mars.nasa.gov/resources/22201/sounds-of-mars-nasas-insight-senses-martian-wind/?site=insight (accessed 20 November 2019)

NASA. (2020). What is weather like on other planets. Available at: https://spaceplace.nasa.gov/weather-on-other-planets/en/ (accessed 30 April 2020)

New World Climate. (2020). How the sun affects weather. Available at: http://www.nwclimate.org/guides/sun-affects-on-weather-tutorial/ (accessed 3 February 2020)

Peters, K (2012). Manipulating material hydro-worlds: rethinking human and more-than-human relationality through offshore radio piracy. *Environment and Planning A*, 44(5), 1241–1254.

Peters, K, Steinberg, P, and Stratford, E (eds.) (2018). *Territory beyond Terra*. London: Rowman & Littlefield.

Rabin, S (2004). Nicolaus Copernicus. *The Stanford Encyclopedia of Philosophy*, Available at: https://plato.stanford.edu/entries/copernicus/ (accessed 13 August 2019)

Sheller, M and Urry, J (2006). The new mobilities paradigm. *Environment and Planning A*, 38(2), 207–226.

Simpson, P (2019). Elemental mobilities: atmospheres, matter and cycling amid the weather-world. *Social and Cultural Geography*, 20(8), 1050–1069.

Sharp, D (2017) What is sound. From Open Learn at the Open University, Available at: https://www.open.edu/openlearn/science-maths-technology/science/physics-and-astronomy/physics/what-sound (accessed 23 September 2019)

Spector, S, Higham, JE and Doering, A (2017). Beyond the biosphere: tourism, outer space, and sustainability. *Tourism Recreation Research*, 42(3), 273–283.

UCAR Centre for Science Education (2020). What is weather? Available at: https://scied.ucar.edu/learning-zone/how-weather-works/weather (accessed 15 April 2020).

Vannini, P and Taggart, J (2015). Solar energy, bad weather days, and the temporalities of slower homes. *Cultural Geographies*, 22(4), 637–657.

Weather Online (2020). Troposphere. Available at: https://www.weatheronline.co.uk/reports/wxfacts/Troposphere.htm (accessed 15 April 2020)

Weir, A (2014). *The Martian*. London: Ray Press.

Whittaker, GR and Peters, K (2021). Research with sound: an audio guide. In von Benzon N, Holton M, Wilkinson C and Wilkinson S (eds.), *Creative Methods for Human Geographers*. London: Sage. pp. 129–140

Yusoff, K (2018). *A Billion Black Anthropocenes or None*. Minneapolis: University of Minnesota Press.

Index

Note: *Italic* page numbers refer to figures and page numbers followed by "n" denote endnotes.

Printed in the United States
By Bookmasters